Low attainers
in mathematics 5–16:
policies and practices in schools

SCHOOLS COUNCIL WORKING PAPER 72

Low attainers in mathematics 5–16: policies and practices in schools

BRENDA DENVIR   CHRIS STOLZ
MARGARET BROWN

Methuen Educational

First published 1982 for the Schools Council
160 Great Portland Street, London W1N 6LL
by Methuen Educational
11 New Fetter Lane, London EC4P 4EE

Filmset by
Northumberland Press Ltd,
Gateshead, Tyne and Wear
Printed in Great Britain by
Richard Clay (The Chaucer Press) Ltd
Bungay, Suffolk

British Library Cataloguing in Publication Data

Denvir, Brenda
Low attainers in mathematics 5–16 – (Schools Council
working papers; 72 ISSN 0553–1668).
1. Mathematics–Study and teaching
2. Slow learning children–England
I. Title   II. Stolz, Chris   III. Brown, Margaret
IV. Series
371.92'6

0047739

ISBN 0–423–51020–7

Contents

5

Foreword

Discussions about 'slow learners' are mentioned in the minutes of Schools Council Mathematics Committee in 1975. At that time Dr Wilf Brennan's work on slow learners was in progress and it seemed better to defer more specific work in mathematics until his report (Schools Council Working Paper 63, *Curricular Needs of Slow Learners*) had been published.

The Low Attainers in Mathematics Project began in 1978; the title was chosen because the limited attainment which a teacher observes in a pupil may be the result of a variety of causes, either singly or in combination, and correspondingly, a variety of responses is required of the teacher.

Two books have been produced by the project team. The first, this report, gives an account of schools' responses to the problems presented by mathematically low-attaining pupils. It leaves no doubt that although much good work is being done today there is an urgent need for further work in the fields of research, curriculum development and in-service training to assist teachers to meet the needs of these pupils. The second book (Holt, Rinehart and Winston, 1983) is an account of relevant research into the teaching of mathematics and will be of interest to all who teach the subject. The two offer much information and suggest fruitful approaches; they will be useful to teachers, advisers and those concerned with in-service training.

The steering committee of the project gave freely of their time to advise, discuss and comment, and to visit schools in the company of members of the project team. It is impossible to speak too highly of the concern, energy and commitment of Brenda Denvir and Chris Stolz, ably supported by Margaret Brown, the director of the project. Indeed, the dedication displayed by all concerned reflects our belief that this subject ranks as one of the most important of all for school mathematics. But we are conscious that the work done is only a beginning. It must be followed up!

JOHN HERSEE
Chairman, Schools Council Mathematics Committee

Acknowledgements

The project team and steering committee wish to acknowledge the co-operation of all those who willingly gave so much help in carrying out the project and in compiling this report.

We should like to thank those advisers who replied to our initial enquiry, helped to arrange visits and provided valuable information. We are also grateful to those who received and helped us on our visits to project schools.

Most of all we should like to thank headteachers and teachers in the schools that we visited, both formally and informally, for their willingness to devote time and effort to ensure that our visits were useful and informative; and not least, of course, the many pupils who talked to us about their work.

Finally we should like to thank the project secretary, Isobel Provan, without whose calmness, efficiency and support this report would not have been completed.

Introduction

The Low Attainers in Mathematics Project arose from the desire of the Schools Council Mathematics Committee to assist those responsible for the teaching of mathematics to low attainers. The need was highlighted by the results of the study carried out by W. K. Brennan for the Schools Council, *Curricular Needs of Slow Learners.*[1] Whilst Brennan expressed concern about the quality of the curriculum in general for slow learners, it was in mathematics that he judged there to be a particularly dismal picture: '... successful curricula [for slow learners in mathematics] were observed in only one in four primary, one in six secondary and one in eight special schools' (p. 100).

What were the aims of the project?

The Mathematics Committee of the Schools Council set up a working party to decide what should be done. Members of the working party visited a number of schools and consulted people who were active in this field. They concluded that work was needed in several areas, namely:

teaching and learning;
selection, diagnosis, screening and evaluation;
school organization;
courses and curricula;
teacher training;
teaching materials.

There was considerable discussion about whether this should be under-taken at the primary as well as the secondary level of education. It was felt that, although major problems became more apparent at secondary school, pupils who are low attainers in mathematics need help much earlier and, hence, the problem should be studied across the statutory age range from 5 to 16 years. To present a task of reasonable scale which could be tackled in a limited time, it was decided to exclude pupils in special schools and nursery schools, and students over the age of 16, whether in school or college.

The working party decided that initially work would be undertaken in the following areas:

1   to bring to the notice of teachers current knowledge about learning processes in mathematics; and
2   to provide information to teachers and advisers about the ways in which primary, middle and secondary schools deal with pupils who have difficulty in mathematics.

The work would be supervised by a steering committee which would also be responsible for formulating proposals for further work required.

These two parts to the Low Attainers in Mathematics Project were concurrent and carried out by members of the same group working at Chelsea College, University of London. The literature research team reviewed relevant research literature whilst the general project team gathered information about current practice in educating low attainers in mathematics. The report of the former, which is still in preparation, will describe and review the research which relates to the teaching of mathematics to low attainers and the difficulties which such children have in learning the subject. The report will be divided into four sections: Number, Measurement, Spatial thinking, and Language – words and symbols. It will include an extensive bibliography.

This book is the report of the general project team. The work ran for eighteen months, from May 1979 to November 1980. It was carried out by two full-time team members and an honorary director, with the active support of the steering committee.

The original intention was for this group to identify good practice in the teaching of low attainers in mathematics in a few schools and then to write detailed case studies of those schools, following the pattern of the Schools Council's report, *Mixed-ability Teaching in Mathematics*.[2] All the schools visited were indeed giving a great deal of thought to the problem and had generally developed many effective practices in aspects of teaching low attainers in mathematics. However, in the judgement of the project team, none of the schools had yet been completely successful in evolving and implementing an *overall* policy for the teaching of low attainers in mathematics including suitable courses, organization, teaching approaches and assessment procedures, to the extent that it could be used as a model for other schools to copy. The plan was therefore revised. Consequently, this report is divided into chapters each dealing with a different aspect of the problem and illustrated with examples from the schools visited.

Since many of the aspects are closely interrelated the divisions are somewhat artificial, but it was necessary to impose a simplifying structure. It is not, however, truly possible to separate, for example, organization and teaching approach.

What is the scope of this report?

Essentially, the project team has looked at current practice in a number of schools in an attempt to identify what was felt to be good and to highlight those aspects which are important. The team has attempted to examine some of the ways in which general school practices affect low attainers in mathematics. There is no simple solution to the problem, and this report does not set out to offer one. Indeed, it may be said in criticism that it offers nothing new. Much of what is here has been said before, but most of it as isolated and separate points.

However, there are several ways in which it is hoped that this report will be of value:

1 It collects together many points of discussion in the hope that authorities, advisers, schools, departments and individual teachers will consider them in relation to their own circumstances.
2 It indicates who or what might help.
3 It suggests, mainly by reference to examples from the schools which were visited, some ways in which present practice might be changed.
4 It identifies some further work that is needed.

For whom was the book written?

The book is intended for those who teach or are concerned with the teaching of low attainers in mathematics. It may be especially helpful for inspectors and advisers, for heads, deputies, heads of mathematics and remedial departments at secondary schools and teachers with responsibility for mathematics in primary schools. It is envisaged that it will be used primarily in in-service training, both inside and outside of school, as a document which will encourage discussion and lead to the clarification of ideas and the formation of a consistent overall policy. Nevertheless, it is hoped that it will also be helpful and supportive to the individual teacher who feels isolated in his or her concern about the teaching of mathematics to low attainers.

Primary and secondary schools may seem very different, but increasingly, as the team members visited schools and examined the apparent effects of current practice, they became aware that many points applied to a greater or lesser degree right across the 5–16 age range. For this reason the reader is strongly urged to consider examples which relate to age groups other than those which are his or her major concern. Where a point applies to primary or secondary pupils only, this is stated.

Inevitably, much of the report has implications for all pupils, not just for low attainers. However the implications are especially significant for low attainers since without favourable conditions these pupils are likely to make little or no progress in their mathematics.

How was the work done?

A number of schools were visited formally by the project team, using a procedure described below. All except two of the visits were made by at least one project team member and one member of the steering committee. In addition many schools were visited informally; information contained in this report arose from both types of visits.

In order to select those schools to be visited formally, the Schools Council wrote to all local education authorities (LEAs) in England and Wales asking for recommendations of schools which might be of value to the project. From the replies, schools were chosen that varied in age range, size, geographical location, type of catchment area, organization, mathematics courses and so on. Table 1 gives the number of schools visited formally in each age range.

Table 1 Schools visited formally by the project team, by age range and category

Age range	Number	Category	Number
3–8	1		
4–7	1		
5–7	2	Infant and first	5
5–9	1		
4–11	2	JMI[a]	2
7–11	6	Junior	6
8–12	2	Middle	2
11–16	1		
11–17	3	Secondary	10
11–18	6		
13–18	3	Upper	3

[a] Junior Mixed and Infant.

Of the secondary and upper schools there were one girls' school, one boys' school and eleven coeducational schools; one was secondary modern, the other twelve comprehensive.

Questionnaires were designed and sent to each of these schools so that the project team could gather much of the simple factual information before the visit. A conscious effort was made to consider all possible factors which affected the low attainers in mathematics. During the visits, the team members talked to many of the teachers who were involved at different levels with low attainers; they also observed lessons and pupils' activities and talked to individual pupils.

As well as visiting schools, members of the project team attended a number of conferences and visited working groups, research projects and advisers. The discussions and visits proved to be of considerable help.

The project team held a weekend conference which was attended by many of the steering committee members and others who were known to have a particular interest and knowledge in this field. The discussions at the conference were of great value in determining the form of the report.

References

1. W. K. Brennan, *Curricular Needs of Slow Learners* (Schools Council Working Paper 63). Evans/Methuen Educational, 1979.
2. Schools Council, *Mixed-ability Teaching in Mathematics*. Evans/Methuen Educational, 1977.

I. Who are the low attainers in mathematics?

Descriptions of low-attaining pupils

Teachers in the schools visited often told us about pupils whom they considered to be low attainers in mathematics. These descriptions illustrate the variety of pupils whom teachers consider to be having particular difficulty in mathematics.

Zynet – an 11-year-old in a suburban primary school. Arrived in England recently; speaks little English; prefers watching to participating but beginning to join in with art and games. Is puzzled in measuring and other practical mathematics. In arithmetic, after initial mistakes, is able to carry out processes to give correct answer.

Brian – a 13-year-old in an inner-urban comprehensive. Good home background; smartly dressed, well-spoken; member of The Boys' Brigade. Enjoys school, even-tempered, well behaved. Works hard but makes slow progress in all subjects, bottom maths set; poor memory. After much work, still cannot remember addition bonds to twenty.

Tracey – a 5-year-old in her second term at a semi-rural infant school. Still cries when mother leaves; hardly notices other children; quiet; little enthusiasm for any activity. Scribbles unevenly and frequently crosses outline when colouring in. Little concentration; poor motor control; handles clay timidly. Doesn't know nursery or number rhymes or sequence of events in simple, oft-repeated stories.

Sharon – 9-year-old in a middle school. Poor reader; has great difficulty learning words and sounds. Art work immature; clumsy in physical education. Does addition and subtraction with numbers up to ten using counters; always counts from one. Lacks 'everyday commonsense', e.g. tries to put paint brushes away in jar which is far too small; difficulty doing up shoelaces.

Winston – 11-year-old in an urban comprehensive. Often in trouble for fighting or insolence; talks in singsong voice, grins; irritates teachers. Makes inappropriate replies. Had remedial reading help at primary school. Enjoys drawing cars, soldiers, guns and strip cartoons and writing war stories. Work generally below standard but especially in mathematics. Has difficulty recording numbers over twenty; does not understand arithmetic operations; can make no sense of structural number apparatus.

Paul – an 8-year-old in an urban primary school. Father died one year ago. Often absent; quiet and withdrawn; incontinent; apparent speech defect attributed to 'immaturity and laziness'. Sucks thumb, daydreams, unpopular, often teased. Enjoys and shows talent in art. Poor reader, handwriting immature, writes *i* for *I*. Long way behind in all number work; has considerable difficulty in recording; generally dislikes mathematics but has shown interest in practical work, shape and symmetry. Responses in spatial work indicate at least average ability.

Jeffrey – a 14-year-old in third year of comprehensive school, organized on mixed-ability lines. Described on transfer from primary as '. . . slow, methodical, independent. Needs extra help with reading. Shows interest in environmental work and practical activities'. Despite enthusiasm for science and humanities, attainment in first and second year poor, owing to difficulty with English and mathematics. Despite poor start, working slowly, seeming puzzled and unaware, he is now making a dramatic improvement in mathematics.

Who do we mean by low attainers in mathematics?

These examples draw attention to the individual nature of pupils' difficulties. Whether low attainers are taught in a group in a mixed-ability class, in a bottom set, or in the remedial department, they will not form a homogeneous group. Their only common characteristic may be low attainment in mathematics.

Many terms have been applied to children who have difficulty with their school work. Among them, 'remedial', 'less able', 'under-achievers', 'slow-learners' and 'disadvantaged' are used in school to refer to children with undefined problems, although the terms were often originally coined to describe children with specific difficulties. The term low attainer is used here because it describes the observable performance of the pupil, without implying a cause.

In recent years two surveys have been carried out which relate to the definition of the pupils with whom this project is concerned. First, the Schools Council Curricular Needs of Slow-learning Pupils Project[1] set out to look at the least successful 15–20 per cent of the school population. However the information collected suggested that it was for a smaller number, between 11 and 15 per cent of the ordinary school population, that schools provide or would like to provide a different or modified curriculum.

The second of the surveys, whose work is set out in the Warnock Report[2] suggests that: '. . . at any one time about one child in six is likely to require some form of special educational provision' (p. 40).

These surveys supported the steering committee in making the decision that this present study *would be concerned primarily with those pupils outside special schools who fall, for whatever reason, into the bottom 20 per cent of mathematical attainment in their age group taken over the country as a whole.*

In secondary schools which set or stream pupils, this definition will generally include pupils who are in the bottom set and also many from the sets immediately above it. In classes organized on mixed-ability lines, the low attainers in mathematics typically form a small group who are well behind the rest of their class and seem to make little progress.

The proportion of low attainers in mathematics in a school as defined above will depend on the school's circumstances. Schools in affluent areas may have few pupils in the lowest 20 per cent of mathematical attainment taken nationally, whilst in some schools in deprived areas a majority of the pupils may come into this category. Sometimes, because these latter schools adjust their expectations to match the majority of pupils, only a small group of the lowest attainers may be thought to cause particular concern. It is nevertheless hoped that this report will be found helpful in every school which is conscious of problems in dealing with a group of children considered to be low attaining in mathematics, whether or not the individual members of this group fit the definition given above.

Many of the recommendations for the education of low attainers in mathematics can be regarded as applying generally to the teaching of mathematics and therefore have implications for all pupils, especially those who at any level are under-achieving or have difficulty in coping with the mathematics that is presented to them. But although the majority of pupils will make some progress without the careful consideration recommended here, the low attainers may not only fail to learn but could leave school at sixteen apparently achieving no more than they had several years before.

What are the causes of low attainment?

Teachers and advisers consulted by the project team quoted a variety of reasons for low attainment in mathematics, for example:

'The pupils are culturally deprived. Their language is limited and they get very little encouragement and support at home.'
'They can't grasp relationships.'
'Low intelligence.'
'Many have had insufficient practical experience in the infants'.'
'Too fast a teaching pace in the early years ... later on, teachers with

limited knowledge of the early stages of mathematics and of the subject's development.'
'Pupils get hang-ups about the subject because they keep getting work marked wrong. Then they either switch off or mess around.'

For the individual pupil it is unlikely that one can ever state with certainty what are the causes of his or her low attainment. However it is possible to identify, in general, those factors which are most frequently associated with failure and are thus likely causes of low attainment. Most of these would be covered by the following categories, which are not exclusive:

  physical, physiological or sensory defects;
  emotional or behavioural problems;
  impaired performance due to physical causes, such as
  tiredness, drugs, general ill-health;
  attitude: anxiety, lack of motivation;
  inappropriate teaching;
  too many changes of teachers, lack of continuity;
  general slowness in grasping ideas;
  cultural differences, English not first language;
  impoverished home background;
  difficulty in oral expression or in written work;
  poor reading ability;
  gaps in education, absence from school, frequent transfer from one
  school to another;
  immaturity, late development, youngest in the year group;
  low self-concept leading to lack of self-confidence.

For many low attainers there will be not one, but several of these factors present. There may be any degree of severity and almost any combination of causes so that the nature of each pupil's low attainment will be highly idiosyncratic and possibly unique.

For some pupils, however, there may be one original cause of low attainment, which, if unchecked, is likely to produce a variety of problems. As time passes the cause and its effects intermingle and become difficult to separate. Eventually, the prime cause may be swamped by the secondary difficulties and attempts to help the pupil may be directed at the effects without the initial cause ever being detected. Winston (see individual examples above), in fact, was found in his last year at junior school to have a severe hearing loss in both ears. Deafness may or may not have been the cause of all his difficulties but it was certainly responsible for some of his problems and its discovery led to a change in the teacher's attitude towards him and to a re-examination of the way that he was taught.

The cause of Zynet's low attainment was primarily her lack of English whilst Paul not only had an emotional problem, difficulties with reading and recording, and lacked motivation, but also had gaps in his education because of absence from school and inability to be attentive when he was there.

The causes of low attainment can be subdivided into:

a   those beyond the control of the school;
b   those partly within the control of the school; and
c   those directly within the control of the school.

Causes which are beyond the control of the school include pupils' physical characteristics and home background. The fact that they are beyond the control of the school does not imply that the pupils cannot be helped. Because Paul's teacher knows about his home background and emotional difficulties she is considerate towards him and tries to make his school life a non-stressful, enjoyable experience. A pupil's poor eyesight cannot be remedied by the teacher but checks can be made that a child who suffers in this way wears spectacles and is able to see the blackboard, books and so on, thus minimizing the effect of the difficulty.

Causes which are partly within the control of the school often spring from the interaction between pupil and environment. Parental attitudes, pressures from the media and from the peer group will influence pupils' attitudes; but so will the ethos of the school. Rutter,[3] in a detailed study of twelve inner London secondary schools, found in examining attendance that: 'There were large and statistically significant differences between secondary schools [p. 92] ... the differences between schools ... *were* systematically related to their characteristics as social institutions [p. 178].'

A pupil may become over-anxious as a result of parental attitudes. For example some parents may themselves have experienced difficulty in mathematics and pass on anxiety or even fear of the subject to their children, whilst others pick out mathematics as a 'key subject' and generate anxiety in their child by an over-emphasis on the importance of success, sometimes reinforced by threats of punishment for failure. Early failure may lead later to a rejection of the subject as 'boring and useless' so as to maintain self-respect, and pupils may subsequently refuse to participate in lessons actively. The attitude of the school and the approaches adopted by individual teachers can be critical in such cases. Pupils are influenced for good or ill by their experiences of the subject matter, the way that it is taught and probably, above all, by their relationship with the teacher. In particular, potential truants who do feel that they are learning something useful or interesting in mathematics lessons are more likely to attend than those who do not.

Colleen, a 15-year-old, attended a school which insisted that girls should wear white socks as part of their uniform. She regularly turned up for the mathematics lessons, which she enjoyed, wearing the regulation socks over her preferred blue ones, then removed the white pair before the next lesson, hoping to be sent home.

c) *Causes which are controlled by the school* include:

inappropriate teaching methods or content;
lack of suitable materials;
lack of responsiveness to the pupil's problems or lack of teacher's time to reflect on the pupil's difficulties and plan suitable work;
a teacher's lack of detailed knowledge of the mathematics being taught, including a knowledge of which skills, concepts, etc. are involved;
a teacher's inability to motivate and involve pupils and organize work efficiently.

In-service teacher training within the school or local authority may suggest ways to deal with some of these causes. Effective approaches within the school and classroom will be discussed in later chapters.

Can we identify the dominant reasons for a pupil's low attainment?

The factors associated with low attainment vary in ease of identification. At one extreme, tiredness, poor reading ability and prolonged absence are easily observable whilst, at the other extreme, causes such as neurological dysfunction cannot realistically be determined. Many others fall between these extremes and are detectable but not obvious, for example hearing loss or gaps in early education.

There are three main types of procedure which will assist identification:

a teachers having access to relevant information about the pupil;
b observation of the pupil, talking with him or her and noting behaviour;
c diagnostic testing.

The identification of the causes will not usually lead to a simple remedy but it will increase the teacher's awareness of the pupil's difficulties and hence of the suitability of different approaches and content.

ACCESS TO RELEVANT INFORMATION

As a professional, the teacher should have access to information about the pupil which relates to his or her learning; teachers need to gather and pass on personal details about their pupils. The child who has a hearing loss, who should wear glasses, who lives in a children's home, who has had frequent absences from school or who is taking medically prescribed drugs will usually have these facts recorded on the school profile or record sheets. Teachers should have easy access to these and be encouraged to read them carefully. There should be similar access to relevant information from medical examinations and, where applicable, reports from the educational psychologist or educational welfare officer.

In primary schools, because the typical organization usually means that one teacher has overall responsibility for a class, and because size is frequently small enough for important information to be made generally available, there is little difficulty about dissemination. It may be useful, though, to hold 'continuity meetings' when pupils transfer from one teacher to another. The problem is vastly more complex at secondary school, but the organization can be designed to help overcome the problem of communication. For example:

In an 11–18 comprehensive with 1700 pupils on the roll, a staff meeting was held between 8.45 and 9.00 each morning. General administrative information was passed on at this time but heads of houses were able to use it to hold short meetings of teachers of a particular pupil to inform them of new information concerning that child.

In an inner urban comprehensive with an eight-form intake, heads of year held weekly after-school meetings of form tutors and, where possible and appropriate, some subject teachers, in order to disseminate and receive information so that all could be made aware of pupils' problems. Unfortunately not all staff could attend these meetings and thus their effectiveness was reduced, but the form tutor then became responsible for passing on information to the teachers concerned.

In several of the schools visited each head of year went up through the school with his or her year group and was able to build up a detailed knowledge of the pupils. This assisted the gathering of information but the problem of dissemination remained.

Without formal and regular meetings, information frequently fails to get through to those teachers who most require it. This is especially so if the only available way is to 'send a note'. Teachers are rightly reluctant to commit opinion or hearsay to paper.

Some teachers prefer to find out about the pupils for themselves without being influenced by records. Whilst one has sympathy with this reluctance to adopt other people's assessments at face value, appropriate action for the individual child is only possible if the teachers involved have the necessary information, whether from parents, colleagues or other sources.

Nevertheless it is important that teachers keep their own, or others', judgements under review and avoid 'labelling' pupils since assessments can be mistaken, circumstances may change, and children develop at different rates and in different ways. Many competent adults had emotional, behavioural or learning problems as children which have been overcome with maturation, motivation and a caring environment.

OBSERVATION

Information which will help the teacher to identify a pupil's problems can be gained through observation. This may shed light on the causes of low attainment. Observation may be informal, when the teacher watches the pupil and notes, for example, social responses, ways of working, or reasoning, or it may be more formally planned when the teacher uses a mental or written checklist to guide the observation.

One teacher *in an 11–18 comprehensive* with a new first-year bottom mathematics set introduced a game of snakes and ladders adding the scores of two dice labelled 1 to 6. From this he was able to assess:

whether pupils could match one-to-one;
their facility with simple addition bonds;
the quality of their eye-hand coordination;
their ability to work cooperatively and resolve difficulties;
any flair or creativity in extending the game and developing new rules;
social maturity, for example in being the loser; and
when they cheated, how subtle it was!

In a rural middle school, a group of teachers asked their new first-year class to write a story to illustrate the multiplication sum 9×3. From this, considerable information was quickly obtained about language skills and interests as well as each pupil's understanding of the meaning of multiplication.

Although such activities coupled with informal observation can quickly give insight into many aspects of a pupil's development, their value is limited by the perceptiveness and experience of the teacher as well as by the time available.

Structured observation can be extremely valuable and also helps teachers to develop their informal observational skills. The Classroom Observation Procedure[4] (COP) material, which is concerned with the general development of pupils, includes tasks such as 'drawing yourself' and incorporates checklists.

DIAGNOSTIC TESTING

Written tests are unlikely to give the same sort of information as observation of the pupils but they may provide some indication of pupils' difficulties. For example, mistakes in copying numbers may indicate limitations in short-term memory, and difficulties in setting out computation may indicate poor spatial organization. The more formal tests will, on the whole, indicate that there is low attainment rather than identifying its cause. A further discussion of testing can be found in Chapter VI.

CHECKLIST

Teachers may find the following checklist useful in looking at individual low attainers.

1 Are physical factors, such as deafness, poor eyesight or colour-blindness, contributing to low attainment?

2 Has the child any perceptual difficulties such as reversals, immature motor skills, or poor spatial perception? Are these mild or severe?

3 Does the pupil have a poor memory? Is it poor in the long term or the short term?

4 What is the pupil's attitude towards school, his or her teacher and mathematics? Does the pupil experience an abnormal level of anxiety?

5 Does the pupil have behavioural problems, such as hyperactivity, which contribute to the difficulty?

6 Does the pupil have emotional problems which affect his or her ability to learn? Are these caused by an exceptional home background? Is the situation at home recent or long-standing; temporary or permanent? Are the emotional problems mild or so severe that the child is nearly always preoccupied, finding school and learning irrelevant to his or her situation?

7 Does the pupil experience social difficulties? Is he or she immature in relationships with peers? How does he or she relate to adults, especially parents and teachers?

8 Has the pupil always been considered low attaining or is there a recent drop in attainment?

9 Is the pupil handicapped in mathematics by poor language?

10 Is the pupil handicapped in mathematics by poor reading skills?

11 Is the pupil low attaining in all areas of the curriculum or just in mathematics? Does he or she have any areas of strength?

12 Is the pupil equally low attaining in all aspects of mathematics? If not, what are the areas of relative strength and weakness? For example, is the pupil better at written or mental work? Has he or she a better understanding in number or spatial concepts?

13 Is the low attainment caused by frequent absence from school, changes of teacher or changes of school?

14 Is the pupil suffering from physical discomfort such as hunger or tiredness which makes him or her less capable of learning?

In summary, observation, reflection and discussion, admittedly time-consuming occupations for the busy teacher, are desirable because they will help to identify the problems the pupil is experiencing and hence his or her needs. These activities will result in each pupil being considered as an individual, not just as one of a group of failures all of whom receive very similar treatment.

Points for discussion

1 What opportunity is there for discussion of individual pupils' difficulties and the probable educational implications? Are all the teachers concerned involved? What procedure is there for reviewing information and progress?

2 What are the means of passing on information to new members of staff?

3 What information about low attainers is considered important? How is it collected and communicated?

References

1. The report of the Project is published as W. K. Brennan, *Curricular Needs of Slow Learners* (Schools Council Working Paper 63). Evans/Methuen Educational, 1979.
2. Department of Education and Science, *Special Educational Needs* [The Warnock Report]. HMSO, 1978.
3. M. Rutter *et al.*, *Fifteen Thousand Hours: Secondary Schools and Their Effects on Children.* Open Books, 1979.
4. ILEA Schools' Psychological Service, *Classroom Observation Procedure.* ILEA, 1979. Details available from Hoxton House, 34 Hoxton Street, London N1.

II. Aims and policies

The school's aims and policies for the education of low attainers in mathematics, even when not explicitly stated, are reflected in what goes on in the classroom. Usually classroom practices are considered before aims; it is only in discussion of the practices that the aims become crystallized. An explicit policy for low attainers in mathematics may be set out in a written document or it may be developed through discussion and not written down. Inevitably there are discrepancies between the formulated policy and the classroom practice.

Why teach mathematics to low attainers?

The question 'Why teach mathematics to low attainers?' rightly concerned many of the teachers in the schools visited. A variety of reasons, often presented as 'aims', were offered by school policy documents and by individual teachers. These included:

'To acquire the basic skills necessary for everyday life.'
'To develop the pupil's ability to think logically.'
'To encourage an independent approach to learning by equipping children with the necessary skills of observation, recording, experimenting and the handling and critical appraisal of sources of information.'
'To develop each child to his or her full potential.'
'To provide a wide range of interesting mathematics for pupils of every ability so that each individual enjoys and appreciates at least some aspects of the subject.'
'To use mathematics to increase pupils' understanding of their environment; to be able to interpret "mathematical language" in everyday use and use it to improve communication.'
'To equip pupils with the work skills needed by society.'
'To help them get a job.'

Aims can be at many different levels. They may be stated very generally and apply across subjects; the same aims may be interpreted more specifi-

26

cally in relation to mathematics. If they are narrowed down still further they become specific objectives. For example:

At a general level an aim might be: 'To equip pupils for everyday life'. A mathematical aim contributing to this could be: 'To handle money in buying and selling transactions'.

A related specific objective would be: 'To work out mentally what the change should be if you hand a shopkeeper a note'.

Aims are concerned with both the individual and his or her role in society. At one end of this spectrum is: 'To develop each child to his or her full potential'. At the other extreme there is: 'To equip pupils with the work skills needed by society'. However the majority of aims fall somewhere in between.

Mathematical aims have been set out elsewhere.[1] In general they can be categorized under three broad headings:

Useful: as a tool for the individual and society, e.g. social competence, vocational skills.

Cultural: as a part of our culture of which all pupils should have knowledge and experience.

Pleasurable: as a potential source of enjoyment.

The aims for low attainers are the same as those for other pupils although the priorities will vary depending on the needs of the pupil. If they are to be realistic, the aims emphasized for each low attainer should be related to age, the nature of the low attainment, and his or her particular social needs.

Each of the three categories above is now considered in more detail in relation to low attainers.

USEFUL ASPECTS OF MATHEMATICS

For the adult who is low attaining in mathematics its usefulness as a tool may be limited to minimal social skills. The amount of mathematics needed in daily life is small but some aspects of number, time and money are fundamental.

At work, some jobs require specific mathematical skills although these may be very limited in both range and sophistication. Most jobs, at the least, require employees to plan logically and follow ordered procedures as, for example, in fault-finding and turning-on procedures for machinery.

At school, pupils are helped by their mathematical understanding to make progress in other subjects. At secondary school an understanding of shape is necessary for the interpretation of drawings in the workshop;

the appreciation of population trends in history requires an appreciation of relative magnitudes of large numbers. At primary school, children's early sorting experiences may lead to attempts to classify plant and animal life. This develops skills of sorting and classifying which are also valuable to the adult, for example in using a filing system or organizing storage in a kitchen.

Outside school, facility with mathematical skills can enhance the quality of life; these include the abilities to score in table tennis, handle pocket money and Saturday job money, and plan a holiday which involves obtaining and evaluating information about places, cost and travel.

In considering the needs of society and the individual, it is important to be aware that pupils will be consumers as well as workers. The low attainer in mathematics is open to exploitation by people and by the media, for example, through the misuse of statistics. If unable to manage his or her own affairs the individual may become, to some extent, dependent on society, absorbing resources instead of making a positive contribution. These inadequancies may well be passed on to his or her own children. School is the only place in which many children can learn to extend their ability to choose. The study of mathematics and its applications can give power to the individual to choose wisely and make judgements, giving increased autonomy and thus the ability to break away from the cycle of deprivation and personal inadequacy.

CULTURAL ASPECTS OF MATHEMATICS
All children should be aware of the nature of mathematics and its importance in our culture. This implies that all pupils should experience a broad spectrum of mathematics, but it does not require all pupils to cover the same ground. It also means that pupils should be given some insight into how mathematics is used in commerce and industry.

PLEASURABLE ASPECTS OF MATHEMATICS
If a valid component of the education of the more successful pupil should be the pleasure of discovering the solutions to problems and in aesthetically appreciating patterns in number and form, should this not be equally true for the less successful pupil even though the pleasure may result from work at a simple level?

Two 10-year-olds in a rural primary school were using Cuisenaire rods. They 'made trains' first with the two rods, then the three rods and finally the six rods. The teacher suggested that they line them up alongside each other in the station. David said 'I know. It's because two threes are six!'

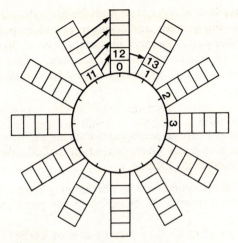

Fig. 1 Glenn's worksheet

A group of 13-year-olds in the bottom set of an inner urban comprehensive school took a sheet on which was drawn a circle, the circumference of which was marked off at twelve regular intervals. They were asked to number the points up to sixty by going round and round the circle (see Figure 1). They were then asked to join the point on the circumference under the number 5 to that under 10, and so on following the numbers in the five-times table. Glenn laboriously and carefully spent most of the lesson numbering the points and joining them up. It wasn't until he'd finished that he said, 'Oh, it's a star!'

The aims for teaching mathematics, as set out above, each indicate a sense in which the pupil can benefit. It is a challenging task for the school to provide appropriate teaching to ensure each pupil does so benefit. But if a pupil in experiences in the classroom is gaining in none of these ways, then there is little justification for his or her continuing. It is for this reason that some people believe that, at least in the last two years of statutory schooling, mathematics should not be compulsory. But it is not possible to say even at this stage that an individual, given such appropriate teaching, will not retrieve the capacity for learning, appreciation and enjoyment.

Needs

LONG-TERM NEEDS
For older secondary pupils the major needs are likely to relate to the

jobs, if any, that pupils may get, their probable life styles and their preferred leisure pursuits. For primary and younger secondary pupils these needs will play a smaller part. The more important need at this earlier stage is not to restrict or hinder potential development. This means keeping options open so as to take advantage of any natural changes or altered circumstances.

SHORT-TERM NEEDS

There are general learning needs pertaining to all low-attaining pupils and specific learning needs which depend on the individual low attainer.
General learning needs include:

a supportive environment and an encouraging reception;
work which is appropriately matched to the pupil's attainment and ability; and
an interesting and motivating presentation of work.

Ideally it is desirable to meet these needs for all pupils, but their importance for learning will depend on the pupil's ability, personality and motivation. A highly intelligent, eager pupil will probably learn even if none of these needs are met, while at the other extreme, for a low attainer who has many difficulties, it may well be necessary to meet all the needs before he or she makes progress.

Specific learning needs for low attainers will be related to the particular causes of their low attainment and may concern organization, teaching approach, course content or materials. For example:

An attempt to meet *Zynet*'s* needs was made by withdrawing her from class for part of each day for special help with spoken English, emphasizing everyday language and explaining important cultural differences.

Paul, who was emotionally disturbed, good at spatial work but poor at recorded work, remained in his class but had work planned for him individually. For instance, he was given a project on wheels which involved a lot of practical activities and led to work across the curriculum. This improved his self-esteem and increased his prestige in the class.

Winston, who was found to have poor hearing, needed early mathematical concepts presented in a way which he found interesting but not condescending, possibly making use of model aeroplanes, soldiers, etc. It was also important that he be addressed clearly, perhaps helped with lip-reading, and that the teacher should check that he had not missed important information.

* Names refers to individuals described in Chapter I, pages 16–17.

Tracey, who was immature and lacking in concentration, needed all-round support at school and, most of all, time to mature and become involved in practical activities and discussion, not hurried too quickly into the more formal aspects of learning or written recording.

Again, a non-reader will need a presentation of work which is not dependent on reading. A pupil of average ability who has been absent from school due to prolonged illness will need extra help in the short term to fill in the particular gaps in his or her knowledge and understanding. For an older secondary pupil who is unable to tell the time and has little understanding of it, a very high priority should be to teach just that.

To determine the priorities for a group of low-attaining pupils it is necessary to consider aims in relation to the needs of the pupils both as a group and individually. A policy translates the required weighting of these aims into classroom practice.

Policies for low attainers in mathematics

IS THERE A 'BEST' POLICY?
A school's policy for low attainers must take account of the school's particular circumstances. These include:

expertise, attitudes and interests of the members of staff;
the needs of the low attainers in mathematics within the school, which may vary according to local circumstances;
the proportion of pupils who are low attainers;
the resources which are available or could be made available in terms of accommodation, equipment (including books), and teaching time.

Although there cannot be one best policy which applies to all schools, some policies will be more effective than others. Successful policies are likely to be those which:

1 take into account the school's particular circumstances as listed above;
2 fit in with any existing overall school policies, in particular those relating to low attainers and to mathematics;
3 are flexible enough to cater for the wide variations in pupils' needs;
4 involve in their development all those who will implement them;
5 take account of available knowledge and resources relating to the teaching and learning of mathematics;
6 can respond to changes in circumstances, e.g. changes in staffing.

These six aspects are discussed or illustrated below.

Policies need to take into account particular school circumstances

In a first school which was in a very deprived area the headmistress felt that all the textbooks available were too difficult, so the staff of the school developed their own materials (see Chapter IV, p. 80).

In another first school one teacher had a particular interest in and knowledge of teaching mathematics to pupils at the early stages of development, so for part of the time she was used as a 'float' teacher to provide an extra member of staff in classes where several pupils were having considerable difficulty with mathematics.

In an area of high unemployment the main local employers set arithmetic tests as a way of selecting future employees. Consequently teachers agreed that arithmetic skills should be given a high priority in the school.

In one open-plan junior school spaces of different sizes were arranged for different activities and teachers were also able to group pupils according to the nature of these activities. This made it easier for pupils who were low attainers in mathematics to receive some teaching in a very small group.

In an 11–18 urban comprehensive nearly all the teachers in the remedial department were ex-primary teachers and had interest and experience in teaching mathematics, so it was agreed that the remedial department would become responsible for teaching the low attainers in mathematics throughout the school.

Policies need to fit in with existing policies, especially those relating to low attainers and mathematics

In one junior school, the new headmaster thought that the standard of arithmetic skills was poor. In an effort to improve standards he administered tests which he had written for all pupils in each year. Staff in the school responded by asking that pupils be set for mathematics, and setting was in fact arranged in three of the four year groups. As a result some teachers felt that low attainers in these year groups became doubly handicapped; they were demoralized by their position in the lowest set and yet were also forced because of the tests to follow the same curriculum as other children, which was clearly inappropriate.

One 11–18 comprehensive was organized on mixed-ability lines because of its aim to give pupils equal status within the school, and every teaching group in each of the first five years catered for the full ability range in that year. In order to cope with the wide range of ability the mathematics department decided to use an individualized scheme which incorporated work for the low attainers. Subsequently teachers found that they did not have enough time to help the low attainers and they arranged for a teacher from the mathematics department to be available in a separate room for some of the mathematics lessons. Those pupils who were having most difficulty with their tasks could go along for help when necessary during these lessons.

In one comprehensive school it was the policy that all 'remedial' pupils be taught all subjects within the remedial department for the first three years. In their fourth year integration of those pupils into teaching groups in the mathematics department caused great problems in both curriculum and teaching approach.

The following framework of policies, whether explicit or implicit, governs decisions at various levels relating to the curriculum, the organization of pupils and teachers, and the allocation of resources.

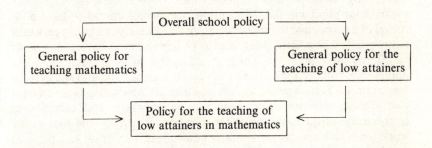

If the needs of low attainers in mathematics cannot be reasonably satisfied within this framework, then policies at a more general level should be reconsidered and perhaps modified.

Policies need to be sufficiently flexible to cater for the number of low attainers in mathematics and the wide variation in their needs
Some pupils have difficulties with mathematics which may, with appropriate teaching and effort on the part of the teacher and pupil, be remedied in the short or medium term. Such pupils' difficulties are primarily associated with such factors as:

frequent absence from school;
inappropriate teaching in the past;
lack of interest in 'education'; or
a background with little intellectual stimulus.

If the teaching of these pupils is successful, it is likely that they will, at a later stage, cease to be low attainers.

Other pupils have difficulties which are, by nature, more permanent. The factors associated with these difficulties include:

low level of intellectual competence;
marked personal and social immaturity;
extreme difficulty in establishing new knowledge or skills;

continually shifting attention and interest; and
extremely poor short-term memory.

Such pupils are likely to need a different curriculum than their peers need for the whole of their school lives.

Whilst it is important to bear in mind that theoretically, at any one time, there are these two groups of low attainers – those who are likely to 'catch up' and those who are not – it is also necessary to recognize that it may not be possible to identify into which of these two categories a particular pupil will fall. In addition, a pupil considered to have difficulties of a permanent nature may mature very quickly at a later age, whilst many pupils whose difficulties seem to be temporary in nature may never be offered the experiences which would enable them to catch up. Some pupils who, initially, make a reasonable start later drop behind and become low attainers. Thus, a policy for low attainers will need to take into account that some pupils will need provision permanently, and others only temporarily; of the latter group the duration, frequency and part of the work for which special provision is needed will vary from one pupil to another.

In one 4–8 first school in a very deprived area there were three special classes. The children in these classes had all seen the educational psychologist and were found to be in need of special help. While some of them were designated educationally subnormal (ESN) and were expected to stay in the special classes for the full four years, others were in these classes because of prolonged absence from school or acute emotional problems, and would return to a mainstream class when their major difficulties had been resolved.

In a girls' secondary modern school the responsibility for low attainers in mathematics was shared by the heads of the mathematics and the remedial departments. Children who had extreme difficulty with mathematics were identified in their first half-term, then withdrawn from their mathematics lessons and taught by the remedial teacher. Girls returned to their own group as soon as possible and by the third year only the very weakest pupils remained in the withdrawal group.

All those who will implement policies should participate in developing them
Policies for low attainers in mathematics may be the decision of an individual or of a group within the school or, in a primary school, developed through general discussion of the whole staff. They may be imposed without consultation or represent a consensus. Sometimes apparently realistic policies are reduced in effectiveness because of the way they were derived.

In a 5–9 first school, the mathematics specialist wrote guidelines which she discussed and agreed with the headmistress and which were then handed on to other members of staff to implement. The other teachers found it hard to identify with the guidelines and to interpret them because they had not been involved in their development.

In one infant school teachers met regularly once a week to discuss their pupils' education. The headmistress encouraged teachers to put forward their own ideas. Each point was discussed in detail and thus the policies were constantly reappraised.

In a secondary school a group of teachers in the mathematics department who were most concerned with low attainers in mathematics, including the head of department, decided as a priority to develop their own materials for low attainers in mathematics. They met two or three times a term to exchange information about which ideas had been successful and to discuss fruitful ideas to explore in the future.

Policies need to take into account available knowledge and resources related to the teaching and learning of mathematics

In a suburban Junior Mixed and Infant (JMI) school a member of the staff was made responsible for the teaching of mathematics throughout the school. She became very interested in the teaching of mathematics; she also became aware that there was much she did not know about the early stages of mathematical development. She asked the head if she could go on a primary mathematics course run by the local education authority. The head agreed on the condition that at some subsequent staff meetings she would outline what she had found out and would initiate a discussion about what implications this should have for school policy.

When he was considering changing the textbook for older pupils, the head of a mathematics department *in an urban comprehensive school* held some of his departmental meetings in the local teachers' centre. This enabled staff to examine and discuss the wide range of books which were on display.

Policies need to respond to changes in circumstances

In some schools policy was decided and the classroom practices determined at some point in the past. Since that time circumstances, including staffing, had changed considerably but the practices had not changed to take account of them. To avoid this situation there needs to be a formal mechanism for regular review of policy at all levels so that a constructive response can be made to changes.

ARE EXTRA RESOURCES NEEDED?

In considering mathematics for the less able pupils in secondary schools Her Majesty's Inspectors (HMI) judged that,

... in nearly one half of the schools [surveyed] the provision for these pupils was unsatisfactory. The schools were more successful in making suitable provision for their average pupils, and still more successful in providing for their more able pupils.[2]

To develop more effective teaching for low attainers in mathematics it may be necessary to allocate more resources to these pupils than are given to other pupils. The extra resources needed may include both teacher time and material resources.

Teacher time

A group of low attainers will probably present a broader range of needs and learning styles than does any other group of pupils. In order to teach them effectively it is necessary to have not only an awareness of the different interests and stages of development of the pupils, but also a detailed understanding of the mathematics which is being taught, because many of the pupils can take only small conceptual steps, and not the greater leaps of their more able peers. On top of this, previous failures and emotional difficulties are likely to lead to behaviour problems, creating further problems for the teacher. Both these factors mean that more teacher time needs to be made available to low attainers than to other pupils. This may be achieved not only by overall increase of the teacher: pupil ratio but also by more flexible allocation of teachers' time.

In an infant school three class teachers had organized their classes so that one of them took all three classes for singing. The two remaining teachers each took out a small group of pupils for extra help with mathematics or language.

Material resources

Low attainers usually progress more slowly than do other pupils, so that, for example, a low attainer may cover in three years the work that most pupils cover in one year. In order not to repeat identical activities the low attainer will need a wider range of materials at each level. In addition, because low attainers usually spend more years of schooling at the early stages of development, where progress depends on handling concrete materials, they will need to have more apparatus available over longer periods of time.

However, this extra cost, even in strictly economic terms, of making suitable provision for low attainers in mathematics when they are at school, may well be smaller than the bill which society eventually pays for its members who are disillusioned or least able to cope.

At local authority and at national level extra resources are also required in such areas as curriculum development, in-service training and basic research. These needs are described in more detail in Chapter VIII.

Points for discussion

1 What are your aims in teaching mathematics to low attainers? Are some of these aims more important than others?
2 What do you consider to be the most pressing needs of the low attainers in mathematics in your school?
3 What is the school's policy for low attainers in mathematics? Does it adequately reflect the aims for, and the needs of, the pupils? Is it realistic and practical? Does it fit in with the general policies for mathematics teaching?

References

1. R. Morris (ed.), *Studies in Mathematics Education*, vol. II: Mathematics Teaching and the Needs of Society. UNESCO, 1981. See also *Why, What and How*. The Mathematical Association, 1976.
2. Department of Education and Science, *Aspects of Secondary Education in England*, a survey by HM Inspectors of Schools. HMSO, 1979, p. 156.

III.   Learning and teaching[1]

The project team observed a wide range of pupils' activities during the school visits. A few examples are given below.

Graham, a 9-year-old in a remedial class, had difficulty with reading and reversed both letters and words, but had less trouble with mathematics. He was playing a game with a friend in the same class. Both boys had an identical set of ten cards, each one portraying a type of vehicle – bus, tractor, motorbike and so on. Previously the children had used these cards in a sequencing game in which the teacher named one, two or three cards and they had to find the correct cards and arrange them in the spoken order.

Left on their own to play with the cards they had extended the game considerably. Each had his cards face upwards on the table. The boys took it in turn to name five cards then shouted 'go' and immediately started to count up to twenty whilst the opponent picked out the named cards and set them down in the correct order. The winner was the first to finish.

When asked if they could do something less noisy than the counting, Graham replied, 'You could use the clock'. They played the game again using the second hand for timing instead of counting.

Ian, a 6-year-old in an infant school, had an exercise book in which his teacher had stamped some clock faces and a work card which said: 'half-past 8', 'half-past 11', etc. He had drawn hands of equal length on the faces to represent these times and for the first one had put:

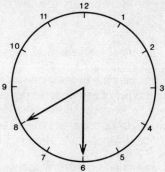

with the hour hand on the eight.

Ian had no idea what time it was (it was about 10 a.m.) and, although he knew that dinner time was at twelve o'clock, when asked whether it was before or after dinner he was not able to reply. When the question was discussed more carefully, he realized what he was being asked, and said: 'Oh, I see. Dinner's coming.'

In the same school as Ian, a 7-year-old girl was working on her own. She had a pile of cards made by her teacher and a one-minute and three-minute sand clock. She chose a card which said: 'Use the 3-minute timer. What can you build with Lego before the sand runs away?' She turned the timer and quickly started building with the Lego. When the sand ran out she took the tower which she had made to her teacher and said: 'I won'. She went back and chose another card: 'Use the 1-minute timer. Walk to the library and back. Were you back before the sand ran away?'

In a comprehensive school, a 12-year-old girl had completed about thirty sums, such as 25

$$\begin{array}{r} 25 \\ \times\ 3 \\ \hline 75 \end{array}$$

The first row of sums in her exercise book were all correct, but the next row began 53

$$\begin{array}{r} 53 \\ \times\ 7 \\ \hline 491 \end{array}$$

and all the rest of the sums had the same mistake of carrying over to the tens on the top line before multiplying. After discussion with the visitor she agreed that she had used two different 'methods' and that the second was wrong.

These illustrations reflect some teachers' assumptions or hypotheses about how children learn, as well as the teachers' philosophy and aims. The examples, however, give little direct evidence of what the pupil is learning. Although one can observe what a pupil is doing, what he or she has learned can only be perceived indirectly in, for example, a change in behaviour or strategy or in the ability to solve a new problem.

What are the different aspects of mathematical learning?

An attempt to analyse the learning of mathematics into different aspects sheds light on the different types of activities that pupils need to experience. It is especially important to be clear about these different aspects when considering the low attainers in mathematics who have already experienced failure. Some of the different aspects are discussed below.

Recall of facts such as simple number bonds like '7 and 5 make 12', and the number of degrees in a right angle.

Acquisition of skills, for example long multiplication and accurate construction of triangles – ordered procedures where questions are posed and answered in standard format.

Understanding of concepts, which can be perceived in the pupils' reasoning. For example, in a second-year junior class:

Two girls had a bag of beans and had to find half of them. One of them described how she had done it: 'We put some of them in each side of the balance. The side that went down, well, we took some out of it and put them in the other side until it was equal. Then it weighed the same, so it was half.'

To take another example, the understanding of a fraction such as $\frac{3}{4}$ involves being able to relate it to a variety of situations, including these represented in Figure 2.

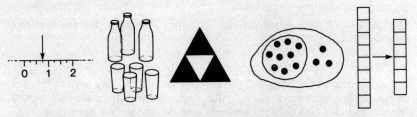

Fig. 2 Pictures which illustrate the concept of three-quarters

Application includes the use of mathematics to solve problems arising in a non-mathematical context. In history, for example, pupils may wish to work out the duration of an historic event from the dates when it began and ended.

Problem-solving strategies, such as drawing helpful diagrams or trying out numerical examples to test a generalization, are ways of tackling a problem when there is no obvious procedure.

Affective aspects include both enjoyment of the subject and an appreciation of its usefulness, power and aesthetic nature.

The categorization of the aspects of mathematical learning summarized above is helpful in outlining the breadth of mathematical education. However it also has a drawback in that it ignores the relationship between the different types of learning. For example, the ease of recalling facts relates to the degree of understanding. Compare the following as answers to '9 × 7'.

'56.'
'I think it's 56, but it could be 63. No, it must be 63 because it can't be even.'
'Hang on a moment, ten sevens are 70, so it must be seven less – 63.'
'Can I get the Unifix?'

The capacity to develop simple problem-solving strategies will depend not only on understanding of concepts, but on affective aspects as well. In order to solve problems pupils need to have confidence in themselves and in the mathematics they have learned so that they are prepared to 'have a go'.

WHAT IS MEANT BY UNDERSTANDING?

'Understanding' involves the recognition of relationships. To distinguish between learning with understanding and learning without understanding, Skemp[2] uses the term 'relational understanding' to mean 'knowing both what to do and why'. The acquisition of skills purely by rote he labels 'instrumental understanding' or 'rules without reasons'. In developing an argument for the advantages of relational over instrumental understanding, Skemp uses the following analogy of the stranger in a town:

> A person with a set of fixed plans can find his way from a certain set of starting points to a certain set of goals. The characteristic of a plan is that it tells him what to do at each choice point: turn right out of the door, go straight on past the church, and so on. But if at any stage he makes a mistake, he will be lost; and he will stay lost if he is not able to retrace his steps and get back on the right path.
>
> In contrast, a person with a mental map of the town has something from which he can produce, when needed, an almost infinite number of plans by which he can guide his steps from any starting point to any finishing point, provided only that both can be imagined on his mental map. And if he does take a wrong turn, he will still know where he is, and thereby be able to correct his mistake without getting lost; even perhaps to learn from it.
>
> The analogy between the foregoing and the learning of mathematics is close. The kind of learning which leads to instrumental mathematics consists of the learning of an increasing number of fixed plans, by which pupils can find their way from particular starting points (the data) to required finishing points (the answers to the questions). The plan tells them what to do at each choice point, as in the concrete example. And as in the concrete example, what has to be done next is determined purely by the local situation. (When you see the post office, turn left. When you have cleared brackets, collect like terms.) There is no awareness of the overall relationship between successive stages, and the final goal. And in both cases, the learner is dependent on outside guidance for learning each new 'way to get there'.

In contrast, learning relational mathematics consists of building up a conceptual structure (schema) from which its possessor can (in principle) produce an unlimited number of plans for getting from any starting point within his schema to any finishing point. (I say 'in principle' because of course some of these paths will be much harder to construct than others.)[3]

The following examples illustrate the difference between instrumental and relational understanding.

Norman, a 7-year-old in a first school, had to do   31   At first, when asked to
                                                    -12

'read' the sum he said: 'Three take away one makes two.' After some questioning he re-interpreted it as thirty-one take away twelve. He was unable to relate this to real life, so he was told the following story: 'There were thirty-one people on a bus. When it got to the stop outside school, twelve people got off, so how many people were still on the bus?'

After hearing this story, Norman was able to work out the answer in his head.

Norman's neighbour was also stuck on the same sum and the teacher showed him how to do it: 'Two from one you can't, so cross out the three and put a two. Two from eleven is nine and one from two is one.'

In a class of 6- and 7-year-olds, the teacher sat with a group of six children who were learning about tens and units. There were several cards on the table, similar in design. Edward, for example, had a card like the one in Figure 3. He clipped tracing paper over the card and drew one cross on the tracing paper over each flower. Then he drew a line enclosing ten crosses. He counted this group again to check that there were ten, then counted those remaining. Next he recorded 5 in the *u* box and 1 in the *t* box (see Figure 4), then handed it to his teacher and she checked it.

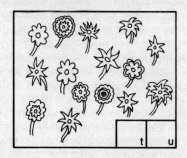

Fig. 3   Edward's card

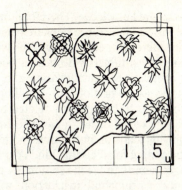

Fig. 4   Edward's completed task

'How many tens?', she asked. 'How many units?' 'How many flowers have I drawn on this card?' Having answered the questions satisfactorily Edward chose his next card which was illustrated with twenty-one birds.

In the first example above the teacher gave a typical 'instrumental' explanation of tens and units subtraction, whilst in the second the aim was clearly for relational understanding.

In Skemp's article [4] he discusses the advantages of relational and instrumental learning:

Instrumental	*Relational*
1 Instrumental mathematics is easier, initially, to acquire. 'If what is wanted is a page of right answers, instrumental mathematics can provide this more quickly and more easily.'	**1** It is more adaptable to new tasks – pupils do not have to learn a new method for each type of problem.
2 'The rewards are more immediate and apparent.' Pupils enjoy a feeling of success when they get a page of right answers.	**2** It is easier to remember, although it is harder to acquire because different relationships have to be recognized. However once relationships are made the learning is more permanent.
3 Pupils can get the right answer to a routine question presented in a familiar way more quickly since they do not have to think it out.	**3** Seeing the relationships is more satisfying than having to learn meaningless procedures, and therefore more motivating.
	4 Each idea grasped is a growth point for further relationships to be developed.

Without forming the relationships which are the essence of relational understanding, a pupil cannot deduce facts or check half-remembered facts, by reference to secure knowledge. Therefore for the pupil who learns instrumentally there will be many more steps to memorize, many of which appear similar but interfere with each other. This will create severe difficulties for those many low attainers who have a poor memory.

The advantages of relational understanding stated above do not, however, imply that the acquisition of skills is unimportant. Rather, they illustrate the need for skills to be explicitly related to concepts which the child has already established in order to assist retention.

The type of learning depends on both pupil and teacher. Skemp points out that each may be aiming for the same kind of understanding or there may be a mismatch. Attempts to teach relationally pupils who have previously received only instrumental instruction will lead, initially, to frustration for the teacher and confusion for the pupils, the latter echoed in the familiar cry: 'I don't want to know why – just tell me how to get the answer.'

The converse mismatch, in which the pupil tries to understand relationally when he or she is taught instrumentally, is also very likely to lead to frustration. This may well be followed by dislike and fear of mathematics and eventually a retreat into demanding only 'instrumental' instructions.

The implications that these ideas have for teaching strategies are discussed in a later section of this chapter ('Can we teach effectively?').

How do children learn?

Surprisingly little is, in reality, known about how children learn but virtually all those who have made studies in this field are agreed that learning is an active rather than passive experience and has its foundation in the child's physical encounter with the real world.

Magne[5] quotes the example of 'Helen – a slow learner' showing how one pupil began to internalize some of her activities:

> Helen was late in grasping the meaning of the letters and numbers. At the age of nine she got a few weekly hours of remedial teaching (in mathematics). Helen did not understand how to write 2 or 3 digit numbers. We let her use a set of blocks we have called the ten-base material and have found most useful for getting low-achieving or under-achieving children to learn the number system.
>
> Helen used these blocks with her exercises. To begin with it appeared nearly impossible for her to realize the meaning of the numbers. Nor did she seem able to see the significance of the material. But an improvement was on its way. At first Helen succeeded in producing the correct set of blocks when the teacher had, for instance, instructed her to represent the number 207, or to tell the number 207 when the teacher laid the blocks in front of her.

For a long time Helen could only work with the help of the blocks. But one day she achieved a dramatic triumph. Helen said to her teacher, "I need no blocks. Because when I write 207, I have two flats and seven cubes inside my head. And no longs at all. And so you see, miss, it is 207." (p. 3)

Piaget believed that activities are vital. He saw pupils as learning from two types of experience. By *physical experience* he meant activity with objects to discover the properties of the objects themselves, for example comparing the masses of two objects using a balance. By *logico-mathematical experience* he meant the child's mental activity relating to the physical activity carried out on the objects. As an example of this he described a child rearranging a set of pebbles, counting each arrangement and so discovering a logical property of number, i.e. that the number remains constant whatever the configuration.

Piaget considered that each child goes through *three stages of development* of this type of logico-mathematical knowledge: 'pre-operational', 'concrete operational' and 'formal operational' (see Table 2).

Table 2 Piaget's three stages of development

Developmental level	Level of understanding of logical ideas
Pre-operational or intuitive	No real understanding; inconsistent reasoning
Concrete operational	Complete understanding at a concrete level; reasoning consistent but always with respect to real objects
Formal operational	Complete understanding at an abstract level; able to reason with reference to logical definitions

A child who is pre-operational, for example, may believe that the length of a stick changes when it is moved. A child who is concrete operational may not be able to make sense of a graph which relates pressure to temperature since the graphical representation is far removed from the physical changes it represents.

Essentially, Piaget maintained that the child cannot learn logical concepts until he or she has reached a certain stage of readiness. He did not, however, suggest that the pupil should do nothing until he or she is ready but that the teacher '... organizes situations that will give rise to curiosity and solution seeking in the child ... Should the child have difficulties in his attempts to grasp a certain idea, the procedure with an active

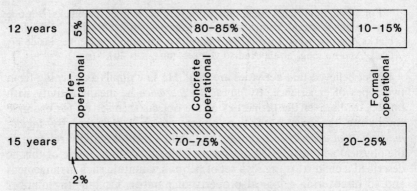

Fig. 5 Distribution of Piagetian stages in British children

Source: M. Shayer, D. E. Küchemann and H. Wylam, 'The distribution of Piagetian stages of thinking in British middle and secondary school children', *British Journal of Educational Psychology*, **46**(2), June 1976, pp. 164–173; and M. Shayer and H. Wylam, 'The distribution of Piagetian stages of thinking in British middle and secondary school children. II – 14- to 16-year-olds and sex differentials', *British Journal of Educational Psychology*, **48**(1), February 1978, pp. 62–70.

methodology would not be directly to correct him, but to suggest such counter examples that the child's new exploration will lead him to correct himself.'[6]

More recent investigations with British children by Shayer, Küchemann and Wylam[7] indicate that the stages of development occur much later in a majority of pupils than had been suggested by Piaget, who maintained that most children were pre-operational from about 1½ to 7 years, concrete operational from 7 to 12 and formal operational from 12 years. For example at 12 and 15 years the approximate percentages of pupils at different stages were as shown in Figure 5.

This figure suggests that 'slow learners' are likely, during the whole course of their school career, to be progressing gradually from pre-operational to early concrete operational in Piagetian terms. For instance, many low-attaining 15-year-olds will not have progressed mathematically beyond the level of an average 9-year-old. The response of two such children,[8] both aged 12, illustrates that using the Piagetian tests[9] they were at the intuitive level in some tasks testing the ideas of horizontal and vertical (see Figures 6–9).

Fig. 6 Yvonne's response when asked to draw trees and a house on the side of a mountain

Fig. 7 Gillian's response when asked to draw trees and a house on the side of a mountain

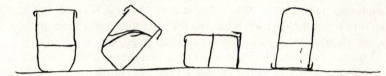

Fig. 8 Yvonne's response when asked to draw the level of water in the jar

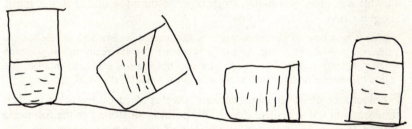

Fig. 9 Gillian's response when asked to draw the level of water in the jar

In the concept testing relating to number operations, Yvonne and Gillian were asked to choose the 'correct sums' to match some simple multiplication and division problems. These, for example, are two of the questions that they were asked:[10]

A bucket holds 8 litres of water. 4 buckets of water are emptied into a bath. How do you work out how many litres of water are in the bath?	8×4 $12 - 8$ $8 + 4$ $4 + 4$	$4 \div 8$ $8 \div 4$ 4×8 $8 - 4$
286 people are coming to see the school play. The chairs are arranged in rows of 13. How do you work out how many rows of chairs are needed?	286×13 $13 \div 286$ $13 + 286$ 13×286	$273 + 13$ $286 - 13$ 22×13 $286 \div 13$

Neither girl was able to choose the correct expressions in these types of questions and this too would suggest that in relation to number concepts both had progressed, at the most, to the early concrete level in Piagetian terms.[11]

Piaget's theory of stages of development is by no means universally accepted. However even his critics suggest that the child's state of knowledge is of paramount importance. For instance, the educational psychologist Ausubel [12] says: 'If I had to reduce all of educational psychology to one principle, I would say this: the most important single factor influencing learning is what the learner already knows. Ascertain this and teach him accordingly.' While this is easy to say it is much less easy to act on since it may involve adjusting the syllabus, or abandoning the textbook where necessary, for each individual pupil!

Low attainers may learn incidentally when they become involved in an absorbing activity in which ideas are presented at their conceptual level, or they may be more conscious of a problem and actively participate in the 'struggle'. In Magne's words,[13] Helen, the girl referred to on page 44 who was able, eventually, to picture the base-ten blocks in her mind, caught a small amount of logic

> after a long time of strenuous work by the teacher ... But first of all the child had exerted herself to the utmost. It is often necessary really to struggle with the problems and to do so with intense concentration and keen interest. But this concentration and interest must be on the side of the child. (p. 3)

Sometimes children learn because they spot inconsistencies in their thinking and try to resolve them. This point is illustrated in the following extract from a conversation which one of the team members had with a pupil in a suburban comprehensive school. Julian, aged 14, had just said that there are fifty tens in five hundred:

TEAM MEMBER   Can you tell me how many tens there are in seven thousand?
JULIAN   Seventy.
TEAM MEMBER   How do you get seventy?
JULIAN   Well, I just guessed.
TEAM MEMBER   Why did you pick on seventy as a guess?
JULIAN   I was going on the first two numbers, I suppose. Fifty was the first two numbers in the five hundred.
TEAM MEMBER   Aha, but you were sure fifty was right. Are you sure seventy is right?
JULIAN   No.
TEAM MEMBER   Why not?
JULIAN   It's a bigger number. It's harder to work out.
TEAM MEMBER   What's a bigger number?

JULIAN   The seven thousand.

TEAM MEMBER   Mm. And so why don't you think seventy is right?

JULIAN   I don't know. It's too small. It's only ...

TEAM MEMBER   Which number is too small?

JULIAN   The seventy, because it's only twenty larger than fifty. While the number you gave me was five hundred and now it's seven thousand. So I think it's wrong.

TEAM MEMBER   Have you any idea what would be right?

JULIAN   No. Seven hundred.

TEAM MEMBER   Seven hundred?

JULIAN   Yeah.

TEAM MEMBER   Why do you say seven hundred?

JULIAN   Cos there're a hundred tens in a thousand and there're seven thousands there.

This illustrates that a situation which provokes conflict in the pupil's mind, in this case when Julian arrives at an answer he feels is of the wrong size, can provide a fruitful trigger for learning.

Can we teach effectively?

Since we do not know exactly how children learn we cannot predict with certainty which teaching techniques will be effective. Nevertheless there are some general principles relevant to the teaching of mathematics to low attainers on which educational psychologists and philosophers seem to agree. Whilst there are some recommendations for the teaching of mathematics to low attainers in the following sections, it is quite unrealistic to suppose that any teacher could put more than a few of these ideas into effect at once. Strategies for introducing more appropriate techniques will be discussed in Chapter VII.

VIEWS OF EDUCATIONAL PSYCHOLOGISTS AND PHILOSOPHERS

The following points have been stressed by educational psychologists in relation to the effective teaching of mathematics.

1   Pupils should have a variety of concrete materials from which to abstract a common mathematical structure.[14]

2   Pupils should be offered many types of apparatus and a variety of approaches because individuals differ in their mental imagery and thus in the ways that they can most easily understand and learn.[15]

3   The teacher should analyse precisely what he or she wishes the learner to be able to do, define what abilities are needed to do it, and then ·take the pupil through a series of subordinate tasks which will help to acquire the abilities.[16]

4 The pupil should have an 'overview' of that which he or she is setting out to learn:

> ... from the beginning you can try to help the child towards some degree of understanding of the general nature of the learning activity that he is engaged in, so that, before he gets down to the confusion of detail, he has at least a rudimentary knowledge of the kind of thing that he is attempting.[17]

Educational philosophers of the past as well as twentieth-century psychologists have commented on teaching strategies. Pólya,[18] the famous mathematician, quotes Socrates with approval: 'The ideas should be born in the student's mind and the teacher should act only as midwife'.

Griffiths and Howson[19] point out that the 'discovery method' favoured by the Nuffield Mathematics Project[20] has its roots in the eighteenth century:

> 'Teach your scholar to observe the phenomena of nature, you will soon raise his curiosity, but if you would have it grow do not be in too great a hurry to satisfy the curiosity. Put the problems before him and let him solve them himself. Let him know nothing because you have told him, but because he has learnt it for himself. Undoubtedly the notions of things thus acquired for oneself are clearer and much more convincing tha[n] those acquired from the teaching of others ... (Rousseau, *Emile*, 1762)'

THE EXPERIMENTAL NATURE OF TEACHING

In the absence of a precise science of learning teachers are really in the role of experimenters, trying out ideas developed by themselves and their colleagues and extending those which are effective. By observing the low attainer and gaining some insight into his or her strengths and weaknesses, present state of knowledge, and probable causes of the low attainment, the teacher can plan work suitable for the pupil. At every stage the teacher can extend, adjust or relinquish the activity, depending on its effectiveness.

Some approaches are tempting to teachers because they appear to be effective in the short term, but often they do not lead to long-term success, or they may lead to success, but only in a very restricted field.

THE TEACHING OF NUMBER SKILLS

Is 'instrumental' teaching effective?
Whilst recognizing that 'understanding' is desirable, many teachers feel that there are facts and skills which pupils should learn by rote even if they do not understand them. The project team met some teachers who felt that for low attainers the objectives should be restricted to learning the rules of arithmetic.

Pupils in the fifth-year bottom set of a comprehensive school were having a class lesson on the addition and subtraction of decimals. The teacher explained how to do the sums, worked through some examples on the blackboard and then asked her pupils to do examples from a textbook, for instance:

$0.6 + 0.35 + 1.27 = \underline{\hspace{1cm}}$.

Find the difference between 1·7 and 3·1.

In the latter type of question two girls repeatedly tried to take the larger number from the smaller.

The teacher said that she had taught the work three weeks ago and that in the meantime the children had forgotten it. Later she added that she would probably have to teach it again since the pupils would not remember.

Many low attainers do appear to have a poor memory for mathematical facts and procedures. This is not surprising as research suggests that there is a reasonable correlation between an individual's short-term memory and his or her developmental stage. As Skemp points out, instrumental understanding relies very heavily on memory because the steps which are to be learned are not related to ideas already in the learner's mind and therefore, particularly for low attainers, instrumental instruction is not likely to prove very effective.

In the Concepts in Secondary Mathematics and Science study,[21] eight of the children interviewed attempted either to multiply or to divide by powers of ten, using long multiplication or long division, which indicated that they did not understand the basis of these processes. Only one of these eight made no errors at all in the three calculations. A typical mistake was made by Tim (13 years) who wrote:

$$\begin{array}{r} 327 \\ \times\, 10 \\ \hline 327 \\ 3270 \\ \hline 3597 \end{array}$$

This suggests that learning the rules without understanding is unlikely to be a reliable strategy.

The project team found that, in some of the secondary schools visited, low attainers in the third and fourth years spent most of their time in the mathematics lessons trying to master the same arithmetic skills which younger low attainers were attempting in the second or third year of junior school. These observations are supported by two recent HMI reports. The report on primary education said: 'Nine out of ten 11-year-old classes were taught to carry out calculations involving the four rules of number to two decimal places or more'.[22] The report on secondary education observed:

Often the teacher's response to weaknesses in calculation was a direct attack on the standard procedures for calculation, believing that pupils would learn simply by watching the processes performed and repeating them afterwards. On some occasions these methods appeared to work satisfactorily, at least in the short term. But such a concentration on calculation, devoid of application and motivation, was often counter productive ...[23]

Thus, on the basis of the observations made during the school visits, it seems likely that some of the children who are unable to master arithmetic skills at the primary stage spend most of their time in mathematics lessons in the secondary school repeating this computation with very little success, for, despite this emphasis on skills, research[24] indicates that pupils show surprisingly little improvement in performance in 'arithmetic skills' over the 12 to 15 age range.

For the subtraction sum

$$\begin{array}{r} 2312 \\ -\ 547 \\ \hline \\ \hline \end{array}$$

the success rates were as follows:

At age	% successful
12 years	61
13 years	62
14 years	62
15 years	66

The implication is that for those children who are unsuccessful at age 12 the teaching of skills in isolation is not effective. Yet sometimes pupils are more capable than would be expected from their performance in written computation. For example, it was reported by the member of the project team whose talk with Julian was quoted above, that Julian could add, mentally, $932 + 4001 + 78 + 6$ but when asked to set it out did not line up the columns correctly.

In marked contrast to pupils' performances in the 'skills' questions of the tests developed by the Concepts in Secondary Mathematics and Science study, the success rates in the questions testing understanding always rose significantly from 12 to 15 years. For example, for the question, 'Write down any number between 4100 and 4200', the success rates were:

At age	% successful
12 years	69
13 years	80
14 years	85
15 years	88

Thus children may still be making errors in carrying out the standard procedures which they have been taught, even though they have developed sufficient understanding about numbers to work things out for themselves if pushed into doing so by appropriate questioning. This implies that the skills which pupils are supposed to have mastered are not always effectively re-examined and re-interpreted in the light of their new understanding, either by their mathematics teacher or by the pupils independently.

How should we teach number skills?
We have seen that pupils are unlikely to remember reliably facts and processes of which they have little understanding. This suggests that the teaching of skills to low attainers can be done most effectively if only those skills which are appropriate to their level of conceptual development are attempted.

Hayley, a 10-year-old, was in the bottom set of the third year in a junior school. She was doing sums from the board and had:

$$898 + 999$$

She worked out the answer correctly using Unifix cubes to find the sum for each column. However, when asked, she did not know which number came after 999 nor did she know which number came after 99.

In the next example,

$$694 - 365$$

she started by taking fourteen Unifix cubes, removing five and recounting the remainder, then writing

$$694 - 365 = 9$$

She then took nine Unifix cubes to work out nine take six from the tens column. When asked if she could do that in her head, Hayley thought for a while, then asked 'Is it four?'

There are obvious gaps in Hayley's knowledge of number which are at a more elementary level than the work that she had been given. Before tackling sums of this complexity it would be advantageous for her to fill some of these gaps, for instance to develop a familiarity with large numbers and how they are recorded. She might, for example, be asked to do a job which involved counting the numbers of each denomination of cardboard coins in the classroom, finding the totals on a number line and recording them each week, then comparing the numbers in a number line to see if any coins are lost. Other work could involve number bonds up to ten, as in the next example:

Richard and Malcolm, two 6-year-olds in an infant class, were shown a game by the teacher. They had pencil and paper, two dice numbered 1–6 and a collection of Unifix cubes. They took it in turns to throw the two dice, add the numbers in their head, write down the sum and take that number of Unifix cubes. When both had done this three times they added up their score by whatever means they wished, recorded the total and then, still using their own method, decided who had got the larger score and by how much. At first Malcolm looked at the scores which they had written down, examined each collection of cubes to decide which was smaller and counted on from the smaller number, picking up a cube each time to record the counting-on process. Richard watched him do this, then said, 'It's easier like this'. He put each collection of cubes in a column, matched the columns and counted the unmatched cubes. The boys then enthusiastically started another game.

This illustrates how the 'recording' of mathematical processes can also be matched to the pupil's stage of development. When the pupil is first introduced to new ideas the work should involve activity with physical objects and discussion. At this stage any recording can be done in terms of physical objects, like the columns of cubes in the example above. Pupils who seem to grasp the ideas at this level may then move on to informal recording, in which they are encouraged to develop their own ways of recording mathematical activities. The standard ways of writing down computational procedures have been developed and condensed over thousands of years and now are short and sophisticated; it is not surprising that children find them difficult to understand. Pupils should only move on to formal recording when they are confident in handling both the ideas and the symbols. Failure at any stage should be followed up by more varied experiences at the previous stage and not by repeated examples at the failed stage.

Only if the teacher succeeds in presenting the work at an appropriate level will most low attainers succeed in learning; frequent success in non-trival activities is vital if the aim for the pupil is to increase self-confidence and reduce anxiety. For pupils who have severe anxiety related to 'getting it right' it may be helpful, for a while, for the pupil to have activities in which there is no possibility of failure, such as self-checking games or puzzles and open-ended investigations.

As Skemp points out, individuals have particular strengths or weaknesses in their perceptual abilities. For this reason the teaching should include a variety of experiences and a wide range of appropriate apparatus and materials. Lessons can be planned to include changes in activity in an attempt to keep pupils alert, responsive and motivated. A proportion of the total lesson time could be allocated to individual work, group work and class work; activities could include mental work, games, puzzles, model making, discussion, recorded work and problem solving.

If only those mathematical ideas which the pupil can grasp are presented to him or her and if only computational procedures which are not too far ahead of his or her understanding are taught, then it follows that not all pupils will encounter all the arithmetic which it is usually accepted that everyone should learn. Whilst this may, at first, seem to be a very unconventional standpoint it is hardly sensible to spend time and effort in teaching the pupil skills which he or she cannot retain for long, and far less apply to problems in textbooks or real life.*

For example a pupil whose understanding of decimals is so poor that he or she cannot read a scale, for example in using a ruler to measure a length of 3·7 cm., can have little practical use for long multiplication of decimals.

One such example of where the skills taught appeared to be beyond the capacity of the child to apply them is given below.

Susan, a fourth-year lower-stream pupil in an urban comprehensive school, told one of the team members that she had recently been 'doing decimals'. The rest of the conversation went as follows:

TEAM MEMBER   What would you use decimals for? Can you think of a situation in which you have to use decimals?

SUSAN   What, add?

TEAM MEMBER   Yes, can you think of a situation, except for a maths lesson, when you have to add decimals?

SUSAN   What, you mean you have to put the point there?

TEAM MEMBER   Yes, not *how* you do it, but *why* you might have to do it. Can

---

* This point is taken up again in the section on calculators in Chapter IV.

you think of any real situation in which you might have to add decimals, or is it just in maths lessons?

SUSAN   Long multiplication you have to do, don't you?

TEAM MEMBER   Yes, you would have to, but can you think of *why* you might be doing it – outside, when you leave school, for instance?

SUSAN   Yeah, they're bigger than money, aren't they?

TEAM MEMBER   OK, ... anything else where you might use decimals apart from money?

SUSAN   Take away.

TEAM MEMBER   Yes. Just in money, that's the only time you'd use them?

SUSAN   Yeah.

DISCUSSING WORK

At every level pupils need help in building up the network of relationships between knowledge, skills and applications. This can be encouraged by posing questions and problems at the appropriate level.

In the conversation with Julian the strategy of getting him to explain his thinking forced him to resolve inconsistencies in his reasoning. That type of questioning strategy will probably only be successful if the questions are at an appropriate level, which is just beyond the pupil's present understanding. Questions which probe children's understanding and encourage them to reason can be put to pupils of any age and are useful for initiating discussion. For example:

Before the child does a sum: 'How big do you think the answer will be?' Or after he or she has done it: 'If that number had been 27 instead of 26, would the answer have been bigger or smaller?'
'Can you think of a story to go with this sum?'
Or, 'Is that a sensible answer? How can you convince me that it is?'

A class, group or individual can be asked, for example, to estimate lengths:

'How tall do you think I am?'
'Is 2·1 metres reasonable for the height of an 11-year-old?'

But only those pupils who are encouraged to give honest opinions because they feel that this is valued by the teacher will be prepared to think and give reasoned replies. Children whose experiences have led them to conclude that only 'the correct answer' is acceptable will not leave themselves open to disapproval or ridicule from either the teacher or other pupils.

Pupils can be stimulated to reason for themselves and develop their self-confidence if they are encouraged to develop their own ways of answering questions and solving problems. Low attainers are often capable of finding the right answer to a question in a less elegant way than the standard method.

In a middle school, one boy aged 11 had to do $102 \div 3$. On a scrap of paper he wrote:

'18	32	36	34
18	32	36	34
18	32	36	34
54	96	108	102'

and then, in his book, '$102 \div 3 = 34$'.

This boy had, in fact, a good grasp of the number operations and a method which is far more versatile than a poorly remembered computational process. In fact adults rarely use formal computation in everyday life, but more often adopt informal methods, such as mentally 'counting on', or practically matching lengths of wood. Children use informal methods too, as reported by the Assessment of Performance Unit Mathematics Monitoring Team:[25]

> In one-to-one practical testing, 11-year-old pupils were given the problem: 'George has 65 stamps in his album and Susan has 102 in hers. How many more stamps has Susan than George?'
> Most pupils worked out the problems in their heads and by far the most frequent method used (by about three-fifths of the pupils) was some form of complementary addition or counting on from the lower of the two numbers to the larger.

PROBLEM SOLVING FOR LOW ATTAINERS

Problem solving is often considered an unsuitable activity for low attainers. Reasons given for this view include:

> they are thought to be unable to think for themselves or make decisions;
> their basic mathematical knowledge is so weak that they will not be able to apply it to the solution of problems;
> there is no time for such a luxury when there are so many skills to be practised.

However there are several reasons why it may be particularly apt:

> the pupils are more likely to understand the mathematics if it is developed in relation to a problem which is itself meaningful;
> pupils are more likely to be motivated to solve problems which have some significance for them;
> problem solving is the way that mathematics is most often encountered in real life;

it is essentially open-ended; pupils are finding out what will work rather than one correct answer so if appropriate problems are selected they can be engaged in activity in which they cannot fail.

Suitable problems may be essentially real-life or purely mathematical and could include, for example:

'How would you get to the swimming pool?'
'How many squares are there on a geoboard?'
'Given three single-digit cards, how many door numbers can you make up?'
'How much should you charge for a cup of coffee in the Youth Club?'
'What could we do to help visitors to the school to find their way round the building?'

Many examples of pupils involved in trying to solve real life problems are described in the Open University 'Mathematics across the Curriculum' course.[26] This also gives many suggestions of suitable problems to tackle in the classroom.

APPLYING MATHEMATICS

If facts and skills which are learned in school are going to be useful outside the classroom then the pupil must be able to apply them to real problems. In emphasizing this, the Schools Council evidence to the Cockroft Committee[27] points out that

> Numerical computation forms but one stage in the solving of a problem. Other stages are evident in the solution of even apparently simple problems and these require differing types of knowledge and skills. (p. 17)

In a simplified account of the processes involved in solving problems of a mathematical nature, described in greater detail by Pólya[28] and Krutetskii,[29] the main steps can be regarded as:

1 Recognizing that there is a problem which could be solved and is worth solving.
2 Formulating the problem and deciding what calculations should be done.
3 Estimating and/or calculating.
4 Interpreting the answers to the calculations and acting on them as necessary.

The Schools Council pamphlet[30] takes as an example:

> ... a joiner's problem of deciding how many pieces, each thirty centimetres long, could be cut from a 5 metre length of wood ... The step of identifying

the structure and representing it in mathematical symbols is known as building a mathematical model ... The next step is to apply mathematical techniques to arrive at a mathematical solution ... The final stage is to interpret the mathematical solution in real-life terms. (pp. 17–18)

The discussion goes on to outline alternative strategies for solving the joiner's problem using mathematical models of various levels of sophistication.

The calculation part, which is the only part of the problem which can easily be done by machine, is usually given a high priority in schools, but there is much less emphasis on the other steps. Some schools, however, are successful in giving their pupils a strong incentive to use mathematics in realistic problem-solving.

In one 13–18 comprehensive school, 15- and 16-year-old pupils who were taught mathematics by the remedial department learned how to interpret timetables, maps and written directions while planning and going on an outing which involved a train journey.

In the third year of a junior school pupils were studying a topic on the measurement of mass. They had started off with a class discussion of why different things, e.g. potatoes, lorries, babies, medicines, needed to be weighed. Then they were asked to write a story or draw a picture portraying an imaginary disaster which had arisen from inaccurate weighing. After weighing and estimating the masses of objects in the classroom in cooking and environmental science as well as in mathematics, the pupils investigated a variety of devices for measuring mass. Later the pupils did both mental and written problems involving mass which were related to their competence in number work.

Because these pupils were given an 'overview' of the idea of mass and were encouraged to develop it across the curriculum, they were helped to see both its usefulness and its place as a part of human knowledge. It is essential that mathematics be related to other curriculum areas. In secondary schools this may involve joint planning between subject departments. Nowhere is this more important than for the low attainers.

In *Aspects of Secondary Education in England* HMI state:

> The difficulty pupils experience in transferring skills and concepts learned in the context of the mathematics lessons to less familiar areas of application is a long standing problem which requires further investigation. Able pupils often manage to make this leap without special assistance but for others there can be serious conflicts in methodology, language and symbolism.[31]

But, in general, they found that: 'Mathematics was too little related to the world outside and to other subjects in the school ...' (p. 159).

Nevertheless some schools make noteworthy attempts to develop mathematical links with other subjects. One example of an attempt to develop mathematics across the curriculum in a secondary school is given below.

In one 11–18 comprehensive school, work was planned and carried out in the mathematics department which would help pupils in science. Activities included, for example, plotting the height of liquid in a bottle against volume for a variety of bottle shapes. Pupils were then asked to explain the shape of the graph and later to predict graphs from bottle shapes and vice versa.

THE PUPIL'S COMMITMENT TO LEARNING

In order to learn the puil has to make an effort and become actively involved. Since many low attainers lack interest in their work, the teacher may have to plan very carefully to create an atmosphere which is conducive to learning and which is likely to increase the pupil's interest and motivation.

The teacher's attitude towards low attainers is of paramount importance. Whilst a sympathetic awareness of a pupil's behavioural, social and learning difficulties is desirable, successful learning is much more likely to occur if the teacher is not patronizing, but able to encourage the pupil's growth towards autonomy and independence. The headmistress of a first school in which many children were disadvantaged described her task as 'extending the boundaries of what makes each child feel safe'.

Pupils respond more readily, too, when they feel that there is a genuine 'partnership in learning' between them and the teacher. An element of independence and choice can be given to pupils, matched to their maturity so that an ability to choose and make decisions is fostered.

In one class of 6- and 7-year-olds in a first school the room was decorated very attractively with children's work which included mathematical displays – symmetry patterns, tessellations, block graphs, etc. One group of pupils of similar ability was working with the teacher on tens and units (this group was described on pages 42 and 43). The rest of the children worked in friendship groups.

Four low-attaining boys were playing 'Beetle'. They each had a card made by their teacher on which a complete beetle was drawn and covered in transparent plastic. The children used paper clips to hold tracing paper over each card and when they had thrown the die they decided what number it was, either by recognition of the dot-pattern or by counting. They then referred to the coding card. On this was written '6 for a body, 5 for an eye', etc. so that they had to recognize the numeral before tracing in the appropriate part of the beetle.

The other rules of the game, e.g. 'You can't put the eyes on before you've got a head' were not written down – the children knew those rules. They enjoyed explaining to the project team member how to play and watched each other carefully to make sure no mistakes were made.

There was a wide variety of apparatus in the room including games, puzzles, sorting activities, structured number apparatus and measuring instruments to which the pupils had easy access.

At the secondary stage some materials have been developed which allow a certain amount of choice on the part of the pupil. The Kent Mathematics Project (KMP) 'L material'[32] gives pupils a free choice for the last of each set of twelve tasks, the remainder being assigned by the teacher. In both SMILE[33] and KMP schemes pupils can choose in which order they will tackle their allotted tasks unless the teacher indicates otherwise.

In a fourth-year remedial class in one 11–18 comprehensive school, pupils were allowed to choose what they wished to do from a bank of material available, although the teacher usually offered some guidance on what he felt was appropriate.

Control

Although this section appears near the end of the chapter, it does not indicate that this aspect of teaching is considered to be of little importance. On the contrary, it is vital that good discipline be maintained if low attainers are to benefit from lessons. Valuable points relating to the general problem of control and discipline of pupils can be found elsewhere;[34] like other general points they also apply to low-attaining pupils.

In the schools visited by the project team the teachers were aware of the difficulties of controlling low attainers, especially the older pupils. Yet in nearly all of the schools the problems seemed to have been minimized because the teachers were well organized and this reduced the occasions of potential confrontation.

In discussions with teachers several points were frequently mentioned as being particularly helpful in the control and discipline of low attainers. These were:

'We try to make our rules clear-cut, consistent and few in number and give reasons for our decisions.'

'I think that pupils need to know what standards of work and behaviour are acceptable and what are not ... low attainers should not be allowed to get away with shoddy work or feel that they are not really expected to do their homework.'

'In our school we try to be caring and concerned about the individual and never give up with any pupil.'

'I try to learn the pupils' names as quickly as possible and make a particular effort to get to know them informally outside lessons. I also try to avoid confrontation: if the atmosphere does get tense, I can sometimes defuse it with a joke.'

'In our department we make sure that new teachers know the discipline and pastoral care structure within the school and make a point of discussing the implications of any changes that are made.'

'I've found that setting work which is too difficult, unclear or even poorly presented can create discipline problems ... I have a reserve of simple work-sheets, games and puzzles which pupils do when they have finished their work or are waiting for help from me.'

'I think good discipline hinges on organization. Pupils are encouraged to collect their own materials and start work as soon as they arrive ... leaving time at the end of a lesson for pupils to check the apparatus and put it away ... It's particularly important to avoid missing lessons with the low attainers.'

'They need to be busy and involved. I talk to the class for short spells only, give pupils a variety of activities and make the work interesting.'

'They respond much better if you show that you are interested and concerned about their learning. I always comment on work, praising effort and good progress and criticizing when I think an improvement could be made.'

Points for discussion

TEACHING APPROACHES AND PRESENTATION OF WORK

1   Does the work take account of the pupil's stage of development and his or her present mathematical attainment?
2   Is the work material-based? Is a variety of apparatus used including mathematical, domestic and natural materials?
3   Do pupils experience a variety of approaches within the lesson and within the course, including mental, practical, oral and written work? Is there an opportunity for individual, class and group work? Do pupils apply mathematics to solve problems which they see as relevant?
4   Do pupils use a variety of methods of recording which are appropriate to the work and to the pupil's stage of development?
5   Do pupils have some choice which is matched to their maturity? Are they encouraged to take responsibility for their own learning?
6   How is the teacher's time spent in the classroom (marking, controlling, directing, informing, observing, questioning, etc.)?

THE CLASSROOM

7   Is the room welcoming with attractive and useful displays which include work done by the low attainers?

8 Are pupils' materials readily accessible? Is the furniture arranged so that pupils can work in a variety of ways?

PUPIL RESPONSE

9 Are the pupils busy, actively involved and concerned about the outcome of their work? Are they successful? Can they talk about what they have done and apply it?

References

1. Much of the content of this chapter is dealt with in greater detail in the Project's literature review, in which each of the areas of basic mathematics is dealt with separately.
2. R. R. Skemp, 'Relational understanding and instrumental understanding', *Mathematics Teaching*, **77**, 1976, 20–26.
3. Ibid., p. 25.
4. Ibid., pp. 23–24.
5. O. Magne, 'The psychology of remedial mathematics', *Didakometry* (Malmo School of Education, Sweden), **10**, 1978, 59.
6. J. Piaget, 'Comments on mathematical education', in A. G. Howson (ed.), *Developments in Mathematical Education: Proceedings of the Second International Congress on Mathematical Education*. Cambridge University Press, 1973, p. 77.
7. M. Shayer, D. E. Küchemann and H. Wylam, 'The distribution of Piagetian stages of thinking in British middle and secondary school children', *British Journal of Educational Psychology*, **46**(2), June 1976, 164–173. Also, M. Shayer and H. Wylam, 'The distribution of Piagetian stages of thinking in British middle and secondary school children. II: 14- to 16-year-olds and sex differentials', *British Journal of Educational Psychology*, **48**(1), February 1978, 62–70.
8. These responses were collected as part of the project, Concepts in Secondary Mathematics and Science, details of which are given in K. M. Hart (ed.), *Children's Understanding of Mathematics 11–16*. John Murray, 1981.
9. K. R. Fogelman, *Piagetian Tests for the Primary School*. NFER Publishing, 1970.
10. These problems have been published in M. L. Brown, *CSMS Number Operations*. NFER Publishing, 1978.
11. J. Piaget, *The Child's Conception of Number*. Routledge & Kegan Paul, 1965.
12. D. P. Ausubel, *Educational Psychology: a Cognitive View*. Holt, Rinehart & Winston, 1968, p. vi.
13. Magne, 'The psychology of remedial mathematics'.
14. Z. P. Dienes, *Building up Mathematics*. Hutchinson, 1960.
15. R. R. Skemp, *Psychology of Learning Mathematics*. Penguin Books, 1971.

16. R. M. Gagne, *The Conditions of Learning*. Holt, Rinehart & Winston, 1977.
17. M. Donaldson, *Children's Minds*. Fontana, 1978, p. 100.
18. G. Pólya, 'As I read them', in Howson (ed.), *Developments in Mathematical Education*.
19. H. B. Griffiths and A. G. Howson, *Mathematics: Society and Curricula*. Cambridge University Press, 1974, p. 78.
20. Nuffield Mathematics Project, *Teachers' Guides* series. John Murray and W. & R. Chambers for the Nuffield Foundation, 1967–74.
21. Hart (ed.), *Children's Understanding of Mathematics 11–16*.
22. Department of Education and Science, *Primary Education in England*, a survey by HM Inspectors of Schools. HMSO, 1978, p. 55.
23. Department of Education and Science, *Aspects of Secondary Education in England*, a survey by HM Inspectors of Schools. HMSO, 1979, p. 139.
24. Hart (ed.), *Children's Understanding of Mathematics 11–16*.
25. Department of Education and Science, Assessment of Performance Unit, *Mathematical Development: Primary Survey Report No. 1*. HMSO, 1980, p. 95.
26. Open University, Course PME 233, *Mathematics Across the Curriculum*. Open University, 1980.
27. Issued as Schools Council Pamphlet 17, *Mathematics Teaching in Schools: Evidence from the Schools Council to the Committee of Inquiry into the Teaching of Mathematics in Schools*. Schools Council, 1980.
28. G. Pólya, *How to Solve it: a New Aspect of Mathematical Method*. Princeton University Press, 1971.
29. V. A. Krutetskii, *The Psychology of Mathematical Ability in Schoolchildren* (trans. J. Teller). University of Chicago Press, 1976.
30. Schools Council, *Mathematics Teaching in Schools*.
31. Department of Education and Science, *Aspects of Secondary Education in England*, p. 149.
32. *The Kent Mathematics Project* (L material). Schools Council, 1979. Distributed by Ward Lock Educational.
33. SMILE (Secondary Mathematics Individualized Learning Experiment). Inner London Education Authority, 1972– . Materials from ILEA Learning Materials Service, Highbury Station Road, London N1 1SB.
34. M. Marland, *The Craft of the Classroom: a Survival Guide*. Heinemann Educational, 1975. Also, P. Francis, *Beyond Control: a Study of Discipline in the Comprehensive School*. George Allen & Unwin, 1975.

IV. Curriculum and materials

Courses

The school visits made by the project team revealed a variety of curricula for low attainers in mathematics at all age levels. Here, as elsewhere in the book, examples are included to illustrate points, rather than as models for other schools to follow.

In one 8–12 middle school the mathematics specialist had written a very broad scheme of work which was divided into four sections:

'number system, computation and money;
measurement of length, weight, capacity and time;
shapes and geometrical properties; and
pictorial and graphical representation'.

Within each section were listed the order in which concepts should be introduced, the mathematical language involved and references to a variety of resources. It was stressed that for pupils of all abilities all sections were equally important; thus low attainers would be offered the same breadth in their mathematics curriculum as the other pupils.

In a 5–9 first school pupils in the third and fourth years were divided into four sets in each year by ability, for their mathematics lessons only. The teacher of the third-year bottom set (7- and 8-year-olds) felt that it was important to concentrate on social mathematics such as shopping, counting change and telling the time.

A small comprehensive had in the past used the Scottish Mathematics Group series of textbooks *Modern Mathematics for Schools*[1] with all pupils but a new head of department considered the books unsuitable for many children because they included too much algebra and geometry and put insufficient emphasis on arithmetic skills. So the majority of first- and second-year pupils spent half their lessons following the textbook and half on extra computation; low attainers spent all their time doing arithmetic.

The principal aim of the teacher of the *third-year bottom mathematics set in an inner urban comprehensive* was to restore the pupils' self-confidence. Pupils usually worked in pairs and had some freedom of choice in selecting work. Activities were chosen from the Schools Council Mathematics for the Majority Continuation Project *Travel* and *Buildings* packs,[2] and tasks which had been written by the teacher. The pupils were encouraged to use calculators and to discuss and plan what they were doing and speculate about the outcomes. This approach and content differed greatly from that for other boys in the same year who were taught in a traditional teacher-directed way following courses leading to public examinations.

APPROACHES TO DEVELOPING CURRICULA

Teachers individually or acting as a group have considerable freedom in determining curricula for low attainers, although some immediate restrictions are placed upon them such as the availability of texts and materials and, particularly in the secondary school, examination syllabuses. However, in the long term these are within the control of teachers and in any school the curriculum for low attainers and the way that this is described will reflect the aims of the teachers whether or not these have been stated explicitly.

In most of the schools visited teachers had been given guidance on the work that was to be covered with low attainers. Curricula were specified essentially in one of the following four ways or in some combination of them:

1 specified in terms of mathematical content
2 defined by a test or examination
3 defined by a textbook
4 specified by pupils' activities often relating to the use of specific pieces of apparatus and material.

Examples of each of these ways are given below.

Curricula specified only in terms of mathematical content

In the junior department of a JMI school which was streamed, the mathematics specialist had written a syllabus for each of the four years, subdivided into 'number' and 'topic' mathematics. The topic mathematics dealt with time, length, mass, volume and capacity, shape and area. Recognizing that it would be impossible to cover all the items, she marked with an asterisk those that were considered most important in both number and topic. It was recommended that all the mainstream pupils should cover the priority items but that low attainers should only cover the most basic of these. Each teacher was then expected to select what seemed appropriate for his or her pupils from the rest of the syllabus, although the low attainers were expected to concentrate on the number rather than on the topic work.

In an inner urban comprehensive school (in a designated 'Social Priority Area'), in which pupils were set for mathematics, the mathematics syllabus for the first, second and third years was described in a 'Guide for work to be covered by pupils'. It was divided into descriptions of work for each of the three terms and these were then subdivided for each year. The work for each term covered three or four major areas of mathematics, for example sets, number and measure. Under each of these headings specific topics were listed. For example, for the first-year autumn term, 'element, subset, union, intersection' and so on were included under the heading 'sets'. This work was incorporated and extended in the autumn term of subsequent years.

Low attainers generally covered work from all the main areas recommended for each term but did fewer topics and in less detail. By the end of the third year many of the topics, even some listed for first-year pupils, had not been covered by the low attainers.

Curricula defined by test or examination

In the junior school (Chapter II, p. 32) where three of the four years were set for mathematics and where the main objective of the teaching was the acquisition of computational skills, the six sets in each year attempted to pursue the same course because at the end of the year all the children in a given age group took the same test.

In an 11–18 comprehensive, mathematics teachers of the first and second years were given copies of previous end-of-term tests instead of a syllabus. The tests indicated both the topics to be covered and the level to which they should be taken.

In another secondary school, the department had, in consultation with local employers, developed a leavers' 'mathematics profile'. The component tests of the profile, which were mainly concerned with computational skills, were used as a guide as to what was taught to low attainers in mathematics.

Curricula defined by a textbook or work scheme

Examples of schools visited where this method is used, together with a discussion of the possible implications, are given later in this chapter.

Curricula which specify pupil activities as well as content

In an infant school, in order to develop children's basic number concepts such as conservation of number, matching and ordering, teachers were recommended to make use of particular materials which were in the school and to use these first in general play, accompanied by questioning and discussion, and later in specific activities initiated by the teacher, such as:

sorting dolls' house furniture and placing in an appropriate room;
putting away the dressing-up clothes by placing each garment on a coathanger;
comparison of the heights of towers built with bricks or Lego.

The mathematics curriculum *in a JMI school* not only indicated mathematical content but also included and stressed creative and investigative work in the subject. One such specified activity, in which pupils aged 10 and 11 were asked to divide equilateral triangles into three parts in many different ways, was observed in progress during the visit of the team. The pupils' initial drawings were simple but enthusiasm to produce more intricate designs led them into questioning and discussion about linear measure, area, angles, rotational symmetry and tessellation.

In a 13–18 urban comprehensive with a large proportion of ethnic minority pupils, the remedial department was responsible for teaching mathematics to the bottom two sets in each year. The department had written a restricted-grade Mode III CSE course because they felt that some of the pupils could probably gain a certificate. The course incorporated those aspects of mathematics which teachers believed would be of greatest value to the pupils. The emphasis was on the use and interpretation of numbers and information rather than on proficiency in written calculations. The course included planning and going on country walks, visiting unfamiliar towns, youth hostelling and travelling by bus and train, and thus involved the use of maps, timetables, and so on.

Of all the differences between these four ways of specifying the curriculum, the most significant is whether the description is solely in terms of content or whether specific pupils' activities are also included. There are advantages and disadvantages relating to both ways of defining the curriculum for low attainers.

Curricula defined by content only

Possible advantages	Possible disadvantages
Initially makes teachers feel secure because it tells them what pupils should know and what they should and will be doing.	Fairly meaningless if a list such as '4 rules of number; 4 rules of decimals; simple fractions; percentages' etc. is not considerably amplified to include the extent to which each is to be taken.
Comparatively easy to write; it relates simply to the objective 'what pupils should know'.	May lead to topics being studied only in the given order and in isolation.
Relatively easy to assess; tests can legitimately include any selection of questions which match the content.	May lead to the omission of practical activities as they are not specifically recommended.
Leaves the individual teacher with freedom to teach in his or her preferred way.	If a content list is too long, teachers may opt for short-term goals and

instrumental instruction in order to 'get through' as much as possible, so that topics can be 'ticked off'.

Creates confusion for pupils if there is a lack of agreement amongst staff with regard to the teaching approach.

Curricula which specify pupil activities

Possible advantages	Possible disadvantages
May lead to a variety of pupil activities including more questioning and discussion.	May make teachers feel that pupils are missing out important facts, skills and topics.
May decrease the possibility of topics being taught in isolation and may increase the likelihood of linking with other subjects.	The purposes of the activities may not be clear to teachers or pupils.
More easily related to activity-based aims, e.g. 'approaching problem-solving in a sensible way'.	May make teachers feel insecure as they appear to have less control over, for example, pupils' activities, questioning and discussion.
Often suitable for pupils of a wide ability and attainment range.	Course is more difficult to plan, write and assess as it needs to be flexible and vary with age and interest of pupil.

Many teachers of low attainers in mathematics indeed found that a content list alone gave them very little assistance in planning lessons. Some felt that they would be greatly helped if they were given guidance in:

1 activities for pupils, such as ordering numbers up to two hundred by finding them on a number line.
2 goals to be achieved as a result of the activities, for example being able to use the index at the back of a book to find information quickly and efficiently.

In a number of secondary schools in particular it was found that for older low attainers, who still had difficulty with early number work and the meaning of the number operations, a fresh look at these ideas arose by considering sophisticated applications of simple mathematics rather than the mathematical idea itself. Thus a syllabus for this age group might consist of:

playing and scoring in table tennis, darts, etc.;

planning and going on a trip to a disco, cinema, skating rink, or football match;

follow-up of any particular interests which pupils may have – football, fishing, motor bikes, cooking, etc. often leading to a study of simple statistics;

planning the purchase of clothes, budgeting, value for money, etc.;

designing the coding on a food or drinks machine.

THE CURRICULUM FOR LOW ATTAINERS IN MATHEMATICS IN RELATION TO THAT FOR OTHER PUPILS

There are several reasons why it is of great importance to construct the curriculum for low attainers carefully and realistically.

In general low attainers will learn less mathematics while they are at school than other pupils will, because they usually require more varied experience to acquire concepts and more frequent reinforcement and application to retain them.

Much of the mathematics that low attainers learn at school is likely to be crucial to their level of autonomy in adult life.

Teachers, especially in primary schools, often claim that the mathematics curriculum is the same for all the pupils. But if some work more slowly they study less than others, and thus the work which they actually cover is bound to be different. There are a number of ways in which the curriculum for low attainers may be related to that for other pupils.

a Low attainers may attempt to follow an identical curriculum to that of other pupils but at a slower rate.

One JMI school used the series *Making Sure of Maths*[3] throughout the school, with pupils starting the *Introductory Book* in the top infants' year. All pupils worked systematically through each book, missing out very little and having to correct most of the mistakes which they made.

b Low attainers may cover the same topics as other children, leaving each topic either when it gets too difficult or when the rest of the class has finished.

In the fourth-year mixed-ability junior class of a JMI, pupils used the Scottish Primary Mathematics Group[4] scheme, working through each topic as a class. Able children and quick workers were given extension work when they had finished until most of the pupils had also completed the work; the class then moved on to a new topic. The two or three low attainers attempted the same work and usually managed to do the initial tasks but then spent the remaining time struggling with activities which were still in the early part of the topic.

A very different example of planning easier work for low attainers within a class topic was found in a *mixed-ability second-year junior class*. In the first lesson of a topic on fractions all the pupils had been introduced to the idea of $\frac{1}{2}$, some to $\frac{1}{4}$ and a few to $\frac{1}{8}$. This had been done by folding, cutting and colouring paper shapes. In the second lesson the teacher divided the class into groups, each containing pupils of roughly similar ability. Each group had different tasks and within the groups pupils worked either individually or in pairs.

The pupils in one group had coloured gummed paper and were making a variety of shapes and cutting them in half in as many ways as they could find.

Another group was using Scottish Primary Mathematics Group's *Stage One Workbook*[5] in which the work involved colouring in one half of a variety of shapes, and deciding whether or not shapes which were printed in the book had been divided correctly into halves.

These two groups consisted of pupils who had done only halves in the first lesson. The other groups were experimenting and finding a half of: a ball of Plasticine, a bag of beans, a jar of buttons, a jug of water, etc., and were using capacity measures, weighing scales, etc., to do this.

The teacher explained that the more able children would do more experiments, then move on to more formal work, whilst the low attainers would spend longer on the experiments and do just a little recorded work.

In an 11–18 comprehensive, first- and second-year pupils spent about half their week's mathematics lessons on topics. Each new topic was introduced by the teacher to the whole mixed-ability class, who then studied it for about a fortnight, working through a series of worksheets written by members of the department. At the end of the fortnight the majority of pupils had usually completed most of the work but some low attainers had only been able to complete the introductory sheets.

c Low attainers may follow a modified curriculum, missing out some of the topics. The topics which are not studied are selected because they are seen as:

too difficult;
less important for low attainers as they are unlikely to study the topics later on;
less important in comparison with other topics in developing social skills and getting a job; or
unlikely to be interesting or stimulating.

In the streamed J M I school which divided its syllabus into 'number' and 'topic' mathematics and marked some items as priorities which should not be omitted (p. 66), the mathematics specialist had written a modified course for pupils in the remedial streams. The modified course assumed that the 7-year-old low-attaining pupils would start at the same level as the main stream pupils, but would proceed through the 'number' work at a slower rate. It was recommended that pupils in remedial classes should spend less time on the 'topic' mathematics.

   d Low attainers may follow a modified curriculum so that they cover the same breadth of topics as other pupils but go less deeply into each. This is exemplified by the 8–12 middle school in which mathematics was divided into four sections, each regarded as being equally important for all pupils (p. 65).

   e Low attainers may follow a specifically designed curriculum, separate and different from that for other pupils.

In a 13–18 comprehensive school a major part of the mathematics for pupils in the bottom two sets involved working individually, page by page, through the *Headway Maths*[6] textbook series.

In another secondary school, an 11–18 comprehensive, the social studies department had developed an integrated social studies course for fourth- and fifth-year low attainers as an alternative to GCE and CSE Mode I courses. It was a three-subject Mode III course incorporating English, arithmetic and social studies. The arithmetic part of the course was concerned with the learning of basic arithmetic skills relevant to the pupil's future life and the work was reinforced by activities such as shopping surveys.

   In the examples of curricula described in sections **a, b, c** and **d** above, the curricula for low attainers have generally been derived from those which are seen to benefit the more able pupils. Consequently many of them reflect the needs of the more successful pupils which may not be the same as the needs of the low attainers.

   For the pupils at the JMI school (p. 70) who worked systematically through the textbook series, the books effectively were the curriculum. However, the series was aimed at a particular age group, range of attainment, ability and speed and way of working. These may have been well matched to some or even the majority of pupils in the school, but for most of the low attainers there were sections, corresponding to their particular difficulties in mathematics, which required a higher level of understanding and a greater facility with reading and recording mathematics than they actually had. Some of the work in these textbooks was suitable for a low attainer. But it seemed inappropriate that such pupils who needed, essentially, varied practical activities, discussion and experience of a variety of methods of recording their work, should be working mainly individually through a textbook and being required to spend much of their time trying to read the question and formally record their work.

In addition, because much of the work was too difficult, the low attainers had little success in that which they attempted and were regarded by other pupils and by themselves as failures, not only because they made a lot of mistakes, but also because they were on a lower level book in the one series of textbooks used.

In the fourth-year mixed-ability class in the J M I school (p. 70) where pupils worked through each topic as a class and most pupils completed the allotted work, the low attainers had a lack of success and feeling of failure similar to that of those in the previous school. They were also very frustrated because they never actually finished anything. In the next example of the second-year junior class the teacher wished the whole class to be studying the same topic at the same time because it increased the opportunity for discussion. She resolved the problem of making the topic relevant to all her pupils by allowing her less able pupils to start off the topic more slowly, spend more time on practical activities, record their work informally and cover fewer stages in the topic. Thus low attainers finished the work that was expected of them at about the same time as did the more able pupils.

The decision of the J M I school (p. 71) to divide its syllabus into 'number' and 'topic' mathematics and to recommend that pupils in remedial classes spend more time on arithmetic than mainstream pupils did, had two consequences. First, the arithmetic was harder than the parallel topic work so the pupils were less likely to be successful. Second, it seemed to be less motivating, involving less practical work and more repetitive work, which required formal written recording, so it was much more likely that the pupils would become bored.

There are in fact potential disadvantages inherent in each way that the curriculum for low attainers is related to that for other pupils. For example, if a low attainer makes good progress but has missed out some topics, then both teacher and pupil are likely to resist his or her transfer to the mainstream because of the gaps in knowledge. However if the curriculum is designed so that low attainers experience the same range of topics but at a lower level than their more able peers in a 'watered down' version of the main course, low attainers may study little which is useful, interesting or relevant. This aspect is highlighted in the H M I report[7] and supported by some of the team's visits, where low-attaining secondary pupils were seen studying topics, such as matrices, in which they were restricted to the acquisition of low-level skills such as addition. In fact, some of these pupils had no idea what matrices were or what you could use them for. Another example was where pupils were working on the topic 'angles', but low attainers did not progress beyond calculating the third angle of a triangle given the other two. In this case, some pupils could not see why it was 'wrong' to measure the angle in the diagram provided. Others could not remember whether they were to subtract from 90°, 180° or 360° as they seemed to have little feeling for the size of these angles.

The assumption that all low-attaining pupils are necessarily slow workers may lead to inappropriate courses. This attitude is typified by courses which expect the able pupil also to be a fast worker and to do all the given tasks, while the low attainer in the same amount of time is only expected to complete the introductory tasks.

Recently there has been discussion about a 'common core' curriculum in mathematics but there has been little agreement on what it should include. An attempt to implement a common core might give rise to several problems for low attainers in mathematics. First, whatever curriculum is chosen it will be too difficult for some low-attaining pupils. Second, even if a very simple core curriculum were adopted, which it was felt almost all pupils could achieve, there would be the danger that this would be interpreted as a maximum rather than a minimum. Low-attaining pupils, and possibly others also, might thus not be offered any mathematics beyond a narrow range of 'skills'.

One suggestion which would overcome these problems is to group together those mathematical achievements which are considered to be of central importance in a series of 'levels' according to their difficulty.[8] For instance in the area of ratio and proportion the first level might include only those applications which involve the ability to double and halve. Increasing a quantity by a half might be included in level two, while operating with ratios such as $3 : 5$ might not be reached until level four or five.[9] This series of levels would avoid the danger of a fixed core being used either to limit or to over-extend the curriculum for particular pupils.

The description of such a core or series of levels would require careful consideration in order to be at the right level of detail: neither so vague as to be meaningless, e.g. 'developing a feeling for number' nor so precise as to lead to an extensive and therefore overwhelming list, e.g. 'being able to subtract two-digit numbers by decomposition'. If it were to be expressed as a list of routine techniques, this would also encourage a narrow concentration on instrumental teaching. However, a core which includes reference to applications and other processes as well as merely learned techniques, and which contains a broad spectrum of mathematical topics could act to improve the curriculum and make it more relevant to the needs of many low attainers.

MATCHING THE COURSE TO THE PUPILS' NEEDS

Curricula for low attainers which are designed around pupil activities may encourage practical work using a variety of materials, discussion of work and application to realistic problem solving. Some points, however, need careful consideration:

1 Before they can generalize from what they learn, low attainers on the whole need a greater variety of experiences of the same sort of activity than do more able pupils. It is usually necessary for parallels between experiences to be clearly drawn rather than merely implied. For example, work in different number bases will not automatically illuminate ideas of place value in the tens system. Better understanding is only likely to be gained after careful discussion of the underlying principles.

2 Low attainers are sometimes less easily stimulated and more reluctant to become involved in an investigation or problem than their more able peers. If their enthusiasm is not fired they are unlikely to learn very much, so it is important to choose and present activities which are likely to be relevant and motivating, and to be prepared to abandon unfruitful ideas.

3 Great care also needs to be taken in the selection of activities, if the low attainer is not to end up being occupied only in carrying out low-level tasks and skills well within his or her existing capabilities. This is especially so if the activity is to be carried out by a mixed group of pupils. Low attainers should be involved in activities which encourage higher-level thought processes such as organization, generalization and developing and testing hypotheses.

An appropriate balance in the curriculum for low attainers will depend on the age of the pupils.

Younger low attainers

Pupils at the primary stage and in the early stages of secondary school, whilst needing to be recognized as low attainers in order to be given appropriate help, may be late developers who, given suitable teaching, will make reasonable progress later on and would then benefit from a course designed for more able pupils. Thus it is important that curricula for young low-attaining pupils are not so restrictive as to preclude this possibility. There are several other points which support a broad-based mathematics curriculum for younger pupils:

1 One branch of mathematics can, and often does, reinforce another. For example simple number work is likely to arise when studying two- and three-dimensional shapes.

2 All pupils should be introduced to a wide range of mathematics.

3 It is more enjoyable and more motivating to study a variety of topics at a suitable level, than to plod away at a restricted course.

4 When the curriculum is restricted it is usually dominated by arithmetic with a sprinkling of work on money and telling the time. In the early stages these areas seem to involve more difficult concepts than do other mathematical topics such as shape or logic.

These points suggest that younger low attainers would benefit from following a curriculum broadly similar to that for other pupils but that they should progress through it more slowly and have more experiences leading to each idea.

Older low attainers

The problem of devising suitable curricula and matching courses to pupils, or vice versa, is much more pronounced in the secondary school, especially for older pupils. In many schools, public examinations determine not only what pupils are taught but also how they are taught, and the pressure to get as many pupils as possible to achieve grades at O level and CSE influences, often detrimentally, the experience of all pupils including the low attainers.

The head of department of a comprehensive had chosen a traditional A-level GCE syllabus because he had felt that it was the most acceptable for university entrance. An O-level syllabus was selected which gave a good grounding for the A level. The CSE Mode I was chosen because it had maximum content overlap with the O-level course, allowing a mixed O-level/CSE mathematics set and a late decision about which of these two examinations pupils should sit. Similar reasoning applied to the development of the school's CSE restricted-grade Mode III examination which was pursued by all pupils in the bottom two sets, almost all of whom were subsequently entered for the examination. Thus a 'watered-down' academic course was offered to the low attainers which was inappropriate in many of the respects already mentioned. Furthermore the teaching approach was restricted almost totally to one in which pupils were shown an example on the board, followed by an exercise for them to do on their own.

Whether, as in the above example, low-attaining pupils sit a public examination, aim for the school's own leavers' 'Mathematics Certificate' or, as in many schools, follow a specially designed course which does not lead to certification, there is likely at this age to be an increase in the pupils' sense of failure and frustration. This may often be accompanied by a sharp increase in absenteeism, which in turn decreases the pupils' chances of success.

Some schools had partially overcome the lack of motivation and problems resulting from irregular pupil attendance by relying less on class teaching and rejecting approaches which involved a gradual development of mathematical ideas. Instead, self-contained mathematical activities were chosen because they would motivate the pupils; whilst used essentially as 'one-off' lessons they could be extended for pupils who attend regularly. This approach could be used for all the mathematics lessons, or some time each week could be allotted to working individually through a textbook aimed at less able older pupils. Some schools achieved a similar, flexible approach by using an individualized learning scheme.

The awareness of parents, teachers and pupils of the difficulties which lie ahead for the low attainer on leaving school, especially those associated with trying to find a job, has led to an increasing number of schools developing 'social mathematics' courses, often in association with other departments. Selecting the content of such a course can, however, present several difficulties. First, since the mathematics studied will probably only be useful if the pupil is able to apply it, skills and concepts need to be developed in relation to the particular practical problems which the pupil is likely to encounter. This raises the second difficulty, that of predicting which problems involving mathematics the pupil will need to solve. Whilst some of these may be anticipated, it is clearly impossible to foresee precisely what demands will be made by society in the future. An effort has to be made on the part of the teacher to understand the pupil's probable life style if such work is to be realistic and at all motivating to the pupil. Work on mortgage repayments and calculating returns on investments in stocks and shares is less likely to have relevance than, for example, using local maps, interpreting diagrams from do-it-yourself kits, and simple household budgeting.

Two general categories of low attainers have been considered above: the older and the younger. More specifically, the particular causes of a pupil's low attainment will suggest his or her needs, as outlined in Chapter II. For example, if a pupil is identified as an extremely slow learner even at an early age it may be desirable to put an emphasis on social 'survival' mathematics, while not ignoring the aesthetic and purely enjoyable aspects of mathematics. Such a curriculum would be unnecessarily restrictive and undesirable for all but a few low attainers.

The development of curricula for low attainers is not merely a question of finding the one 'best' course, but must inevitably be a compromise between the pupil's needs, social pressures and the agreed aims of mathematics teaching in the particular school.

Points for discussion

1 Does the present curriculum meet all the stated aims for teaching mathematics to low attainers, and does it match the desired balance of priorities between the aims?

2 Is it realistic in terms of:

> pupils' present attainments, attitudes and interests;
> time available;
> teaching skill required; and
> materials available?

3 How does the curriculum for low attainers in mathematics relate to their future needs? Can pupils who make good progress readily return to a mainstream course?

4 Is it flexible enough to accommodate the wide range of pupils' abilities, attainments and attitudes? Does it meet the needs of slow learners, pupils who are poorly motivated and pupils who are underachieving through absence?

5 Is it important that there be development and continuity within the course?

6 Does the curriculum cover a wide range of activities, e.g. does it extend and reinforce the pupils' knowledge and ability through realistic applications of mathematics?

Materials

A wide range of materials, both published and teacher-produced, were seen to be used in the teaching of mathematics during the visits to schools. They included:

> textbooks
> topic books and reference books
> workbooks
> workcard schemes and supplementary cards
> worksheets, both commercial and home-made
> prepared sheets, e.g. graph or isometric paper, dots, squares, etc.
> mathematical apparatus; structural apparatus
> electronic calculators

games and puzzles
domestic and natural objects, e.g. measuring jugs, conkers
audio-visual aids, in particular overhead projectors, tapes, slides, film,
TV programmes and posters.

WRITTEN MATERIALS

Textbooks, workcard and worksheet schemes
The recommendations in Chapter III relating to teaching techniques
mention specifically:

the use of apparatus;
discussion and questioning;
opportunities for application and problem solving; and
work at an appropriate level for the individual pupil.

It follows almost inevitably from these recommendations that, whilst some
textbooks and work schemes are undoubtedly more suitable than others
for low attainers, the answer to 'What shall we use for the low attainers?'
can never be totally satisfied simply by selecting a suitable textbook or
work scheme. For the many pupils who have an irregular profile of attainment, that is, those who perform markedly less well in some aspects of
mathematics than in others, any single scheme is likely to require supple-
menting with additional material. The difficulties are even more marked
when dealing with work which has been learnt incorrectly. Nevertheless,
appropriately chosen or carefully developed written material can and
should support the teaching.

It became apparent, seeing the wide variety of written materials in use,
that not only are there very varying philosophies and assumptions underlying the different schemes, but the use, value and importance of the same
material varies considerably from one teacher to another, even when they
are in the same school teaching similar groups of pupils.

In one middle school which used *S M P 7–13*,[10] basically a workcard scheme, all
the pupils except those in remedial classes started on the first box and worked
systematically through each section. Pupils rarely missed out any of the cards.

In a junior school there were two standard textbooks series: *Mathematics for
Schools*[11] and *Maths Adventure*.[12] Topics were introduced by the teacher to a
group of children of similar ability sitting round a table. The work was discussed
by the group and often involved children handling mathematical apparatus. Any
recording of the work was done by the teacher or one of the pupils on a large
sheet of paper with a marker pen.

Pupils were given instructions by the teacher about what they should do next.

Sometimes these were verbal, at other times instructions were jotted down. Only later in the topic did the teacher tell the pupils which pages of the textbook had been selected for them to complete.

In an infant school in a very deprived area, teachers felt that the available published textbooks written for infants started at a stage which was too difficult for their pupils. The staff had thus written their own pupils' workbooks. After a period of working with materials and very simple recording, children were given the relevant workbook. If the children had grasped the ideas, then they were able to complete the workbook very quickly. If they had problems in filling it in, they returned to practical and exploratory activities for a longer period.

In one junior school the pupils used *Mathematics for Schools*[13] as a standard textbook, but every time the book directed pupils to do practical work involving weighing, measuring length, making shapes, etc., the teacher told them to omit the activity and go on to the next page.

In an 11–18 mixed comprehensive, first-year pupils were taught in mixed-ability groups. The mathematics curriculum was based on the School Mathematics Project (SMP) lettered books.[14] Topics were introduced to the whole class and were followed up by use of the textbooks and non-consumable work sheets, many of which had examples at three levels of difficulty and had been written by members of the mathematics department.

Selection of suitable written material

In selecting suitable written material, it will first be necessary to determine the role it is to play, whether it will be used in its entirety to form the whole curriculum or whether only parts of it will be used to supplement or enrich an existing curriculum. If the written scheme is to form the complete curriculum for the whole ability range, as in the middle school which used SMP (p. 79), then the points made in the section on courses will be relevant. The following points may also be helpful in deciding which written materials are appropriate:

1 Have all the teachers who are to use the written materials been involved in the choice?
2 Does the material incorporate, or is it compatible with, the balance between practical activities, discussion, mental work and recorded work which is considered appropriate for the pupils?
3 Does the text match the reading ability of the pupils?
4 Is there a reasonable balance between different aspects of mathematics, i.e. number, space, measurement, etc.? Are connections made between these areas?
5 Does it match, or is it compatible with, the preferred balance between individual work, group work and class work?

6 What materials or apparatus does the text recommend? Are these considered suitable and are they available?

7 To what extent are the explanations clear to the pupils? Is there likely to be a need for so much extra support from the teacher in understanding the text that it would be simpler to abandon parts of it entirely?

8 Is the rate at which ideas are introduced too fast for the majority of low attainers?

9 Is there any flexibility in the scheme to allow pupils with different needs to follow different paths through it?

10 Are the activities likely to motivate the pupils? Is there variety in the types of activity? Do they present the pupil with sufficient challenge, even at a simple level?

11 Is the mathematical language which is used at about the right level? Does the text match the teacher's preferences in priorities of introducing mathematical terms?

12 Are pupils generally only expected to repeat a procedure shown in a worked example, or is there a genuine attempt to encourage pupils to think out their own methods? Are the exercises stereotyped or is there a reasonable variety in the questions in each exercise to allow an idea to be approached from different angles?

13 Are the illustrations likely to motivate the pupils? Are they culturally acceptable? Are they at the right level of maturity/sophistication?

14 Is the format likely to motivate the pupils? Is it clearly set out and is the size of print, amount of writing on each page, etc. appropriate?

15 Is there a comprehensive teacher's guide?

16 If the scheme is separate from that used with the majority of pupils, is it compatible with it so that pupils can easily transfer from one set of materials to another?

APPARATUS

In Chapter III it was stated that the use of a variety of concrete materials and mathematical apparatus was of primary importance in the teaching of mathematics to low attainers.

Mathematical apparatus may help the pupils' understanding either because it is a physical model of an abstract idea, as in the case of Dienes multibase blocks, or because it provides a simplified illustration of an idea, where the real-life materials would be more confusing. For example, a group of infants who are asked to sort Logiblocs into 'yellow' and 'not yellow' will be asking themselves, 'Is it yellow or not?' and replying unequivocally 'yes' or 'no'. Asked to sort autumn leaves into 'yellow' or 'not yellow', the same pupils will find difficulty in making some of the decisions because of the variations in the leaf colours. A simplified introduction to a concept will help the child to grasp it, but, in order that such concepts can be applied to reality, the pupil must be gradually weaned from the apparatus, by using more varied objects.

In the same way that children may be able to sort mathematical apparatus but not real-life objects, so children may be bound within a particular model of the number system. In using Cuisenaire rods, for example, $2 + 3 = 5$ can be clearly illustrated to some pupils using red, green and yellow rods. But unless these rods are also related to 'real' number situations, such as number of children in the class, 'how many bottles of milk to-day?', or scores in games, the pupil will be studying Cuisenaire rods, not the number system.

Despite this strong plea for the widespread use of concrete materials in the classroom, simply using apparatus is no more of a panacea than is using the 'correct' textbook. Just as pupils can be instructed to manipulate symbols using paper and pen, they can be persuaded to move pieces of plastic or wood about ritualistically and obtain the right answer, but with little understanding. In most cases the real test is whether the mathematical ideas can be applied correctly to other situations.

The project team found that the use of apparatus in schools varied considerably. In general there was better provision and use in primary schools than in secondary, but there were notable exceptions to this.

In one junior school there was a variety of apparatus available, but in most classes the use of concrete materials was severely restricted. The majority of children used only Unifix cubes and these, not as a matter of course, but only when they were unable to do the sum without them. Consequently pupils regarded it as an admission of failure to use the cubes. It took the teacher in the remedial department a long time to reassure the pupils who went to him for extra help with mathematics that it was perfectly acceptable to use any aids which were at hand to support their work.

In an inner urban 11–18 comprehensive, pupils in the first five years were taught in mixed-ability groups using an essentially individualized, activity-based learning scheme. In each mathematics room there was a wide range of mathematical apparatus, including electronic calculators, which was readily accessible to all pupils. The materials formed a natural and integral part of the learning process. Pupils of all ages and ability used them when it was appropriate.

In an 11–16 coeducational comprehensive where first- and second-year pupils were taught in mixed-ability groups, virtually no mathematical apparatus was used. In one second-year class the pupils had been told to complete an algebra exercise from the textbook. However, three boys were unable to do this work. Two of them had been given instead subtraction sums such as $31 - 17 = \underline{\hphantom{xx}}$. The third was doing even simpler sums and using buttons to help him obtain the correct answers.

In one open-plan infant school, apparatus concerned with shape, number, money and time was available in each home bay. In addition three trolleys containing apparatus related to capacity, length and mass were moved around each half-term so that different groups of pupils could use them. A wide range of apparatus was seen in use during the one-day visit including Unifix, number balances (equaliser), Stern rods, number tracks, number squares, sorting apparatus, sand clocks, money games and various shapes and sizes of containers for capacity work.

In one junior school the mathematics specialist had identified what he felt was a particular problem in the school – that pupils had little understanding of place value. In an attempt to improve this he took groups of children from different classes and spent a great deal of time on this topic using Dienes multibase materials.

Selecting apparatus
In selecting apparatus to support the teaching of a particular topic, several points need to be considered.

First, in general it is valuable to have a variety of materials in order that pupils have different experiences from which they can generalize; commercial or teacher-made apparatus ought to be supplemented with household and natural objects.

Second, some pieces of apparatus are of restricted application and essentially have only one use, such as a trundle wheel, while others, centicubes for instance, can be used in a great variety of ways.

In a first school visited by the project team, pupils working individually or in groups used Cuisenaire rods (structural number apparatus) to make very complex, three-dimensional symmetry patterns. Pupils then recorded these, some of them drawing a plan-view and side and front elevations.

Third, the purpose and design of the apparatus should match, broadly, the maturity and sophistication of the pupils. In particular, the size, colour, fabric and the motor skills needed for its manipulation may well indicate that certain apparatus will be more successful with one age group than another. Brightly coloured plastic, always an attraction to infants and younger juniors, may be rejected as babyish by older secondary pupils. Dienes wooden multibase arithmetic blocks (MAB), useful at both junior and secondary stages, may appear drab to most infants and in addition may be difficult for them to handle because of the size of the unit cube. Many children are also likely to experience frustration in attempting to join centicubes. On the other hand, pegboards, and to a lesser extent geoboards, might be used through almost the entire age range, if appropriate tasks are set.

Finally, the apparatus, especially if it is structural, needs to be at an appropriate conceptual level for the pupil. For example, the purpose of the hoop abacus is to represent physically the standard procedures of addition and subtraction. It does little to help the child who has no understanding of place value, who is more likely to be assisted by multibase blocks.

It is desirable that some apparatus appropriate to the age, maturity and attainment of the pupils (including measuring equipment and, in primary schools, objects such as for counting and grouping, and number line) be permanently available and easily accessible in every classroom. Other more specific pieces of apparatus could be pooled and shared by several classes.

Working with apparatus

Apparatus can serve several functions. For example, it can extend the pupil's understanding and enrich his or her learning. When introduced to a new piece of apparatus, pupils should initially be allowed time to explore and 'play' with the materials. The 'free play' stage is later accompanied by discussion and questioning: 'What have you done?', 'Tell me about the shapes' and so on. During these activities, observation by the teacher is extremely valuable in increasing his or her awareness of the pupil's present knowledge. Only after these initial stages is it desirable to move on to direct questioning and initiating specific activities.

In one classroom a 7-year-old boy had been given a balance, some Plasticine and a box of various objects. He seemed to be putting things in the balance pans more or less at random. In discussion, it emerged that he had not linked the idea of heavier and lighter with the movement of the pans, although he appeared to understand the words.

Apparatus can serve to focus the pupil's attention on important points. Apparatus sometimes helps to avoid elaborate preparation for work such as the drawing of diagrams and preparation of tables, which distracts the pupil from the main points of the work. For example:

In investigating the shapes of triangles pupils can use geostrips to make a large number of triangles quickly and accurately.

Apparatus can be used in recording. At all but the initial stages of handling materials, pupils should be encouraged to record their work, but it is important that they see clearly the relationship between their activities and the recording. Initially the apparatus itself can be used for the recording by mounting it semi-permanently on card. Drawing round the apparatus when it is needed for another purpose would lead naturally to a pictorial representation. In the final stages, before moving away from the apparatus to a more symbolic form, it will be necessary to link the manipulation of the materials with the form of recording the pupil is expected to use in the future.

Apparatus can be helpful in consolidation. For low attainers especially, once a particular level of understanding has been reached it is highly desirable that necessary facts (e.g. 9 and 3 are 12, there are 24 hours in a day) be learned, and that some skills (e.g. accurate measurement with a ruler, converting from a circular clock to a digital one) be practised until they become automatic. This implies the need for frequent repetition. Pupils are more motivated if the necessary practice is embedded in even a simple game or puzzle. Such activities have the added advantage that they encourage mental work and are frequently self-checking.

Finally, apparatus can be used in reinforcement and recreational activity. Sensible use of Altair design sheets, mazes, numbered dot-to-dot patterns and so on can be enjoyable and educative, particularly if regarded not as a one-off activity but as part of a continuing thread, and used and referred to for example in discussion or displays. These may also be helpful in releasing the teacher to attend to other groups.

Games and their selection

The use of appropriate games is of particular value not only because they can be highly motivating and provide an opportunity for pupils to practise and consolidate their knowledge, but in many cases they also create a learning situation.

Most of the comments made concerning selection of mathematical apparatus also pertain to the selection or design of games. In addition, if the game is to be successful the following points are important:

rules must be simple, unambiguous, and appear to be fair;
the game should result in a clear winner;
the game should have an element of luck built into it;
the game must be flexible enough to allow it to be played at various levels of difficulty;
the game should actively involve pupils and encourage them to verbalize their answers;
it is helpful if the game is self-correcting; and
preferably, games should be relatively short so that a number of repetitions are possible in the time allotted.

One major difficulty is the introduction of new games with fairly complex rules. This problem was largely overcome in one class by the teacher sitting down with a small group of pupils of various abilities and playing the game with them until they had picked it up; she then used them to teach other pupils.

Audio-visual aids

Variety in the way work is presented can be effective in increasing the attention span of pupils. It can also encourage a situation where the presentation of work is more flexible and takes account of pupils' weaknesses (especially with reading) as well as their strengths.

In a number of schools teachers had tape-recorded work for their low-attaining children. Often the pupils were given written material as well as a tape recorder and headphones. This allowed the pupils to be relatively independent, since they were encouraged to play and replay the tape until they had understood. Use of headphones increased pupils' concentration as they were less easily distracted. They also tended to become more self-reliant.

In a large 11–18 rural comprehensive school where mathematics was taught to mixed age and ability groups, the mathematics department had produced its own scheme of work based on individualized learning. The teaching materials consisted of tapes, workcards and booklets which were supplemented by published workbooks and film strips with synchronized tapes. The audio-visual material appeared to be well integrated within the system and used frequently. Children seemed relaxed and able to cope with the machines and the overall organization.

Television and films may also be especially suitable for low attainers. Work presented in this way, provided it is stimulating and at an appropriate level, is likely to be attractive to pupils.

In a suburban comprehensive, 15- and 16-year-olds following a restricted-grade CSE course watched the BBC series 'Everyday Maths'. Their teacher developed the themes using the published teacher's guide[15] and pupils' worksheets.

Electronic calculators

Electronic calculators are considered separately because, since they have only recently become cheap enough to be readily available, their significance and full effect on the teaching of mathematics is yet to be realized, especially their implication for the teaching of low attainers.

Calculators were seen in some primary schools and a number of secondary schools. They were usually of the four-function liquid crystal display type. Attitudes on the use of calculators in the secondary school were wide-ranging.

In an 11–18 comprehensive, the head of department said, 'We don't allow any pupils to use them here. We teach them how to do it. Some of the pupils do, however, use them for homework but we try to discourage them.'

In a different 11–18 comprehensive there were calculators in all the mathematics rooms and pupils were encouraged to use them when they needed to do so. The use of calculators was also required in some of the assignments which formed the course.

A third comprehensive allowed only their most able pupils to use calculators freely. Average-ability pupils aiming for CSE could use them only in certain sections of the syllabus, and the 'less able not at all, because they would never learn if they had a machine to do it for them'.

The results of research in which calculators have been introduced into primary or secondary classrooms indicate very strongly that pupils' mental and written number skills improve, or at worst do not deteriorate, whereas their attitude to mathematics generally becomes more favourable.

The introduction of calculators does not solve all the numeracy problems of low attainers. Rather it helps to emphasize the need for abilities other than computation which are necessary if pupils are to solve problems in the real world. For example, pupils need to:

know when it is sensible to use a calculator and when it is quicker to do a calculation mentally;
understand the number operations $(+ - \times \div)$ and recognize which should be applied in a particular problem;
know in which order to press the buttons, e.g. $24 \div 8$ or $8 \div 24$;
be able to check the answer by: re-formulating the statement and calculating it in a different way; roughly estimating the answer; and judging if the answer is sensible.

It has been suggested that basic numeracy may be defined as 'the ability to use a four-function electronic calculator sensibly'.[16] By removing much of the drudgery of calculating, the sensible use of the calculator should permit pupils to experience a wider range of mathematics and its application as well as make it possible to tackle more realistic problems, for example deciding which can of soup is best value for money.

Some of the 'sum-setting' calculators such as The Little Professor,[17] although limited in use, may be worth considering for improving pupils' recall of number bonds.

Micro-computers
It is likely that in the future micro-computers will be of some value to

low attainers in some aspects of mathematical learning. However little suitable software has yet been developed. At present there are several research projects which are investigating the uses of computers for teaching mathematics in schools, including the Schools Council projects Machine Assisted Teaching and Computers in the Curriculum.

Points for discussion

1 Is there apparatus which will help children to develop the mathematical ideas which are in the course? Is there a variety of apparatus which lead to the same idea? Is the use of apparatus an integral part of the learning experience?
2 Do pupils talk about their activities with apparatus? Are they able to relate the manipulation of the material to their recorded work?
3 Is the apparatus appropriate to pupils' age, ability, manipulative skills, sophistication and so on, for example the complexity of gradations on scales of measuring instruments?
4 Are calculators available? Do pupils use them? Is the course designed with the realistic use of calculators in mind?

References

1. Scottish Mathematics Group, *Modern Mathematics for Schools*. Blackie and W. & R. Chambers, 1969– .
2. Schools Council, *Mathematics for the Majority Continuation Project* (packs 1, 2, 3 and 4). Schofield and Sims, 1974–75.
3. T. F. Watson and T. A. Quinn, *Making Sure of Maths*. Holmes McDougall, 1966–73.
4. Scottish Primary Mathematics Group, *Primary Mathematics: a development through Activity*. Heinemann Educational, 1975–78.
5. Ibid.
6. R. Hollands and H. Moses, *Headway Maths* (Books 1–5). Hart-Davis Educational, 1977–79.
7. Department of Education and Science, *Aspects of Secondary Education in England*, a survey by HM Inspectors of Schools. HMSO, 1979, chapter 7, paragraph 3.27, p. 121.
8. Joint Mathematical Council of the United Kingdom, *Basic Mathematical Skills: Curriculum and Assessment*. Shell Centre for Mathematical Education, University of Nottingham, 1977.
9. Evidence relating to what abilities might be contained in these 'levels' is given in K. M. Hart (ed.), *Children's Understanding of Mathematics 11–16*. John Murray, 1981.
10. School Mathematics Project, *SMP 7–13*. Cambridge University Press, 1977– .

11. H. Fletcher, A. Howell and R. Walker, *Mathematics for Schools*. Addison-Wesley, 1970–71.
12. J. Stanfield, *Maths Adventure*. Evans Brothers, 1971–77.
13. Fletcher, Howell and Walker, *Mathematics for Schools*.
14. School Mathematics Project, *School Mathematics Project: The Main School Course* (Books A–H). Cambridge University Press, 1968–72.
15. *Everyday Maths – Teacher's Notes for Autumn, Spring and Summer Term Programmes*. BBC Publications, 1978–79.
16. M. Girling, 'Towards a definition of basic numeracy', *Mathematics Teaching*, **81**, 1977, 4.
17. The Little Professor. Manufactured by Texas Instruments.

V. Organization

Initially the project team attempted to deal with organization at primary and secondary levels separately. However it became apparent that, while there are in general considerable differences between a typical primary school and a typical secondary school, no major form of organization for teaching mathematics is restricted to either the primary or the secondary stage of education. Furthermore, the effects of similar forms of organization tend to be similar irrespective of the age of the pupils. Consequently organization and its implications are considered together for primary, middle and secondary schools.

The essence of school organization is to plan the use of resources, such as time, space, teachers and materials, so as to optimize the effectiveness of the learning experiences for all pupils. Clearly the organization will reflect the school's priorities in terms of, for instance:

its educational aims;
its beliefs about how children learn and what is appropriate teaching;
its emphasis on particular aspects of the curriculum;
its emphasis on particular age groups within the school; and
its emphasis on particular attainment groups within the school.

Within the overall school organization there will be an organization for teaching mathematics. A rational approach to this will depend on:

the school's philosophy and its priorities;
the pupils' range of attainment, interests, etc.;
the number of teachers available and their particular skills;
the available resources, including teaching materials and classrooms;
the teachers' views of mathematics and mathematical learning; and
in particular the teaching methods which are thought to be most appropriate.

Especially at secondary level, there is frequently an organization for teaching low attainers, for example in a 'remedial class' or a 'compensatory department'. The organization of the school, the organization for teaching mathematics and the organization for teaching low attainers, where it exists, will all affect what happens to the low attainers in mathematics.

What are the implications of the school and subject organization?

During the school visits it became clear that each form of organization had drawbacks as well as advantages. These were often pointed out by the teachers.

In an open-plan infant school children were taught in five vertically grouped classes. Nearly all of the pupils had previously attended the nursery class; consequently the headmistress was able to allocate children so that each class had a fairly even distribution of pupils who were likely to have learning difficulties or behaviour problems.

The headmistress was committed to open-plan education and teachers had come to share this view. They saw several advantages:

a As each of the classes covered the full age and ability range, each teacher had responsibility for a similar group of pupils. Thus many problems were common, and they were shared and discussed.

b The sharing of problems was also facilitated by the open-plan classroom which allowed teachers to observe each other and encouraged honesty between teachers. They could admit not knowing how to tackle a difficulty and discuss it with others without fear of ridicule.

c On the whole teachers wanted to teach mathematics on an individual basis, especially with the low attainers, because they felt that children were more likely to respond in a one-to-one situation and also that this method made it easier to present work at an appropriate level for each child. The openness of the building facilitated this: children who were not the focus of their own teachers' attentions could be seen by several other teachers and this inhibited anti-social behaviour.

d Because of the organization and the individual approach, the low attainers in mathematics were not labelled as a group who were different from other pupils. In fact, although the teachers clearly knew which pupils had difficulty with mathematics, this was one of the few schools visited where project team members could not quickly identify which pupils the school considered to be low attainers in mathematics.

In another infant school, 'rising fives' were admitted in September, but those with birthdays in the second half of the year attended mornings only for one term. It was pointed out to the project team that in the middle infant classes many of the children who had most difficulty with language work, reading and mathematics were these younger pupils with birthdays between March and August. They had a double handicap: not only were they younger and less mature, but their slightly older peers had benefited from having more schooling in the first year and from the additional advantage of being in a very small class in the afternoons.

There were three classes in each of the three years. Every class had pupils from the full ability range and age range for that year. Each class was a little larger than it would have been if all the staff had responsibility for a class; one full-time teacher, one part-time teacher and the headmistress each taught children in small extraction groups. These groups were not specifically for less able or more able pupils, but so that all children, at some time, would have the opportunity for more individual attention in many aspects of the curriculum, including language, reading, writing, mathematics, art, craft, environmental subjects and music. Sometimes more than one group at a time was withdrawn from a class, leaving the teacher herself with only a small group. The teacher with responsibility for mathematics, who also had a particular interest in the low-attaining pupils, often took groups of pupils who were having difficulty with mathematics. Because of her particular interest and expertise she was able to offer these pupils stimulating work at an appropriate level. Since many other groups were also withdrawn from their class at different levels and for a variety of subjects, these pupils did not feel particularly singled out or labelled as different from other children.

At one time pupils in the junior department of a JMI had been allocated to classes strictly by age, and poor readers were withdrawn for extra reading help. It was felt that the extraction system worked well, and it was extended and pupils were withdrawn more frequently. This caused some disruption of lessons and a suggestion was made that such disruption could be avoided by streaming. The idea was discussed, agreed and implemented: pupils were allocated to the 'A' stream, 'B' stream or 'remedial' stream. Infant teachers allocated pupils leaving the infant department at 7 years to the 'A' stream only if they were competent in both English and mathematics, and to the remedial stream if they were very poor readers. Whilst mainstream classes in each year occupied adjacent rooms, remedial classes were grouped together as a partly separate unit.

The remedial teachers each had a special interest and expertise in teaching reading and this was very valuable to the pupils, but there were some problems related to the teaching of mathematics in these classes. Because of the way pupils had been selected, their attainment in mathematics covered a wide range. The teachers perceived that the best way to deal with this range of attainment was to teach pupils either as individuals or in small groups so that work could be matched to the pupils' attainment. This turned out to be impractical because it was not found possible to organize the classes so that most of the children could work sensibly on their own whilst the teacher took a small group. The staff felt that this inability to work responsibly without frequent interjection from the teacher was partly due to their pupils' reading difficulties and partly to the low morale of the remedial pupils who were aware that they were 'different' from the mainstream pupils and conscious of their poor status.

From the observations made in primary and middle schools, it seemed that when pupils were streamed or placed in a 'remedial class' the selection was almost invariably made on reading skill and this led, almost inevitably, to many pupils receiving inappropriate teaching in mathematics.

The headmistress of a first school (*5–9 years*) in which there were many immigrant children, some of whom spoke no English, had been at the school for four years. When she first came she had felt that morale in the school was low. Teachers had been attempting to do individualized work in mixed-ability classes but she felt that this was not successful. In mathematics, for example, children simply worked through a textbook scheme and teachers spent their time chiefly in marking and administration so that there was little time for teaching.

An attempt to overcome these problems was made by setting pupils for English and mathematics in the third and fourth years (7- to 9-year-olds). After a trial period of one year the situation was reconsidered. It was decided to continue setting for mathematics but to stop the setting for English, chiefly because it was felt that:

> it made the timetable too rigid;
> it precluded the development of ideas across the curriculum; and
> it was not an ideal long-term solution: additional support staff used flexibly as float teachers in the classroom would be much better.

The head felt that setting had been a useful step in improving the teaching of mathematics in the school in general, but that she would like to abandon it in the future and use the extra teachers made available by larger class sizes to offer both a withdrawal system and additional help in the classroom.

In an 8–12 middle school, staff used information from the one feeder first school to allocate pupils to classes so that the range of abilities and the number of children with behaviour problems was approximately the same in each class. In the first year (8- to 9-year-olds) and the fourth year (11- to 12-year-olds) each class covered the full ability range but there was a 'remedial class' in each of the second and third years.

Children were selected for the remedial classes primarily because they were poor readers, and these classes were considerably smaller than the mainsteam classes. There was no remedial class in the fourth year because it was felt that pupils should get used to being in a larger class before moving on to the upper school. In the first year there was no remedial class because staff had thought that the number of pupils needing extra help would not justify it.

In the first two years pupils were taught mathematics mainly by their own class teacher. Some pupils in these two years had extra help in extraction groups: from a language specialist in reading and language, and from a compensatory teacher in a variety of subjects, including mathematics. The headmaster spoke of his concern about the widespread use of extraction groups. He felt it was important to arrange that only one or two teachers have responsibility for emotionally insecure pupils, because of the difficulty such pupils have in responding positively to a greater number of teachers.

In the third and fourth years pupils were set for mathematics; the lower the set the smaller was the pupil:teacher ratio. The mathematics specialist wrote tests for third- and fourth-year pupils which were administered at the beginning of the year. The tests were divided into number work and topic work, and the results were used to allocate pupils to sets. Subsequent movement between sets took place at the first half-term when the teachers had had time to judge if any pupil was wrongly setted. After that there was little movement between sets for the rest of the year because this would have led to a lack of continuity for pupils who were moved.

An 11–18 rural coeducational comprehensive was committed to mixed-ability teaching and resource-based learning. On entry, pupils were placed into mixed-ability groups. The most academically advanced pupils, and pupils with learning difficulties, were shared amongst the groups but care was taken in trying to maintain friendship groups. Pupils were taught all subjects in mixed-ability groups except for modern languages. Those with reading difficulties were extracted from this subject for remedial help. In mathematics not only were the teaching groups mixed-ability, they also included pupils from across the whole age range from 11 to 16, using an individualized, resource-based learning scheme. Experienced teachers also had some A-level students in with the others. Pupils in the lower years who were weak in 'basic number' were withdrawn for some of their lessons and taught in smaller groups. It was hoped that eventually a float teacher, who would mainly help the low attainers within the mixed-ability groups, would replace the present withdrawal system.

The organization allowed great flexibility for teachers and pupils: it made it easy to operate a guided option system. For example, pupils in the fourth and fifth years had the option of doing extra mathematics and often it was the pupil who was comparatively weak in the subject who chose to do this. It also allowed the pupil:teacher ratio and the age range within a group to be adjusted according to the experience and the ability of the teachers.

Pupils were allocated to streams on entry to a *13–18 urban comprehensive* on the basis of their overall ability as indicated by their middle school reports. Pupils were reallocated on the results of school tests at the end of the first half-term; after this there were few changes.

Years were divided into four bands related to external examination goals. Streams in the top band followed GCE courses, those in the second a mixture of GCE and CSE courses and those in the third band only CSE. The two streams in the bottom band, which were the permanent responsibility of the remedial department, followed either non-examination courses or the restricted-grade Mode III CSE courses which the remedial department had developed. The school had deliberately made the top streams large, with thirty-two pupils, in order to create smaller-sized remedial forms with about twenty-five pupils in each.

Most subjects, including much of the mathematics, were taught to pupils in their streamed classes. However, older pupils in the middle bands were set for mathematics so that each set could follow a particular course leading to either a GCE or a CSE examination.

The teaching of mathematics to pupils in the remedial classes presented particular problems. Selected because of overall low attainment, pupils in these classes represented a wide range of ability and attainment in mathematics, from innumerate to potential middle CSE grades. In addition there were a number of Asian pupils who had considerable difficulty with spoken and written English. Even with twenty-five pupils in a class, the smaller pupil:teacher ratio did not permit the degree of individualized attention required by pupils with such diverse learning handicaps.

A large 11–18 comprehensive was organized on a faculty basis. Block timetabling of classes allowed each department to choose the teaching arrangements it preferred. About two years previously the school had rejected the option system for fourth and fifth years and now each faculty was expected to provide courses suited to the pupils' abilities and needs. This had greatly simplified the overall school organization. Most departments, but not the mathematics department, taught first-, second- and third-year pupils in mixed-ability groups, while in the fourth and fifth years grouping was more by ability related to the course chosen.

The arrangements for mathematics were very different. The nine classes together with the remedial classes in each year were grouped into two or three equivalent blocks each containing pupils covering the whole range of ability. Blocks in a particular year were timetabled for mathematics at different times. With the exception of the first term of the first year, when they were taught in mixed-ability groups, pupils were set within each block; the lower the set the smaller was the pupil:teacher ratio.

The compensatory (remedial) department taught the bottom set in each block and when pupils seemed able to cope they were transferred to a mathematics set. The mathematics department would have preferred to teach the whole year group at a single time so that the sets would cover a narrower range of ability, making class teaching easier and more effective.

This arrangement, unlike that in the previous example, allowed greater individual attention for the pupils who were felt to be in greatest need. Pupils joined the school at 11 instead of 13 years, and were taught in groups of ten, returning to a mathematics set when they were able to cope. Thus of the thirty pupils initially identified as having special problems with language or mathematics, only ten, who were essentially very slow learning, remained in the classes taught by the remedial department by the beginning of the fourth year.

Thus the form of the school or department organization will affect the teaching of mathematics to low attainers. The possible implications of each of the main forms of organization are summarized below.

MIXED-ABILITY
This is where the range of ability within each mathematics teaching group reflects the full ability range within the year group.

Several variations were found in schools
1   Random or nearly random selection of pupils for classes, e.g. alphabetical.

2 Children allocated to classes so that the range of ability in each class is maximized, e.g. by distributing the most and the least able pupils.

3 Topping and/or tailing: the most and/or least able pupils taught separately for one or many subjects, leaving restricted mixed-ability groups.

Some variations were associated chiefly with primary schools

1 Pupils allocated to classes strictly according to birthday so that classes cover a narrow age range. This tends to limit slightly the range of ability.

2 Family and vertical grouping permits overlapping of different ages. This tends to increase the range of ability.

Possible effects of a mixed-ability organization on low attainers in mathematics

a Children are not socially isolated from their peer group.

b Children are not so readily categorized and labelled ('slow', 'less able', etc.), by either teachers or other children.

c Particularly in secondary schools, mixed-ability grouping is often associated with individualized or small-group teaching, but this is not always the case. The individualized or small-group teaching is usually achieved by following either the same course at different rates or a unique course constructed for each pupil from a material bank.

d Teachers are more likely to perceive and treat children as individuals with differing needs: for example by trying to match the level of work to the pupil, catering for children's irregular attainment profiles or speed of working.

e However, teaching mixed-ability groups is likely to make greater demands on the teacher in terms of: administration; selecting and/or producing learning materials; the need to know pupils in order to provide appropriate work; and record keeping. Because of these pressures, teachers may:

class teach when it is inappropriate, aiming work at the middle of the class and leaving the low attainer unable to cope;

spend most of their time administering and marking and doing little teaching;

use a totally individualized scheme which may inhibit discussion and group work;

be so overwhelmed trying to cope with the range of work and progress that they cannot give sufficient time to the problems of the individual low attainer in mathematics.

f Because not all teaching approaches are appropriate teachers may be required to learn new skills in order for the teaching to be successful, but since several teachers within the school will probably have similar groups they may decide to solve some of their common problems by working together and developing materials. This in turn could lead to greater conformity and agreement about policies.

g Depending on the ethos of the class and the personality of individual pupils, low-attaining pupils who work with more able peers may become dependent on or demoralized by the more able pupils or, on the other hand, they may learn from the more able pupils.

h Whilst this form of organization requires that a range of materials appropriate to the full ability range be available in each teaching base, pupils within each group are likely to be doing different work at a particular time. Thus less duplication of materials is required in each group.

i If mixed-ability classes are blocked for mathematics the system could be flexible allowing:

variation in class size so as to ease introduction of inexperienced teachers, etc.;
movement of pupils so that those with behavioural problems could be split up, minimizing overall disruption, etc.; and
team teaching or use of float teachers.

STREAMING
This is where pupils are allocated to teaching groups for all or most subjects on the basis of 'overall ability'. The overall ability is assessed either in some combination of basic subjects (i.e. language, mathematics, general ability) or, sometimes, in reading ability only.

Some variations were found in schools
1 Topping and tailing: the most able and least able pupils taught separately, whilst the remaining pupils are taught in restricted mixed-ability groups.

Some variations were associated chiefly with secondary schools
1 The year population divided in two, three or more blocks or houses, either randomly or using criteria which aim to produce blocks with approximately the same range of ability. Pupils within each block are streamed and the resulting streams cover a wide range of ability.
2 Banding, so that a year group is divided into two, three or more ability bands. Additional setting may take place within each band.

3   The remedial department taking responsibility for teaching the bottom stream(s).

Possible effects of streaming on low attainers in mathematics

a   A self-fulfilling prophecy may apply, that is, pupils who are identified as a 'low ability' group to their teacher and themselves are likely to do badly simply because that is expected of them.

b   Valid and reliable procedures for assessing ability, attainment and potential have not been devised so it is not possible to be sure that pupils are in the 'correct' stream. Thus all streams are likely to have a wider range than might be expected. In particular, pupils are often allocated predominantly on language ability/attainment. This is a poor predictor of mathematical ability, and therefore there is likely to be a wide spread of mathematical ability in the bottom stream.

c   Pupils who are successful in some aspects of subjects but have severe learning difficulties in others present particular problems in both initial allocation to a stream and subsequent movement between streams.

d   In so far as it is possible to identify a group of pupils at a similar learning stage, the restricted range of ability within a streamed group lends itself to whole-class teaching. The content and language of the lesson can be aimed to suit the middle of the stream. Especially in secondary school, different teaching approaches may be used with the different streams because the teacher is usually left to choose the approach most suitable to him or herself and the particular pupils.

e   Streams can proceed at different rates suited to the majority of pupils and different courses can be selected to suit the steams. At secondary level streams can follow different examination courses.

f   After streaming and initial adjustments there is often little movement of pupils. It is often difficult to move pupils up because they have experienced a restricted curriculum.

g   The pupil:teacher ratio can be adjusted to give small bottom streams.

h   Because pupils in the remedial streams are often socially immature as well as having language problems, there is particular difficulty in providing work for them to do on their own. As most pupils require frequent attention, opportunities and time available for teaching are very limited.

i   The bottom or remedial stream may become a 'sink' with a high proportion of pupils having behavioural problems. This often results in teacher and pupils having low morale. The teaching of bottom streams usually has low status and hence low priority when decisions are made.

j   The bottom or remedial streams are usually taught by a restricted number of teachers thus creating a more secure environment for the low attainer. The most appropriate teacher in terms of experience, qualification and personal qualities can be allocated to a particular stream. For example a teacher who is particularly sympathetic and knowledgeable about learning difficulties may take the remedial group. In practice, however, teachers may also be streamed, the higher streams having the more able, qualified teachers whilst the lower streams have the younger and inexperienced teachers.

k The low attainer when separated from more able peers may be more willing to try and more responsive in discussion.

SETTING

This is where pupils are allocated to mathematics teaching groups by 'mathematical ability'. This may be assessed by published or school test(s) of arithmetic and/or other areas of mathematics, or it may be based on teachers' observations or on some combination of these two.

Some variations were found in schools
1   Pupils in a year set for all mathematics lessons or only for some of them (perhaps for the teaching of arithmetic).

Some variations were usually associated with secondary schools
1   The year population divided into two, three or more blocks or houses either randomly or using criteria to produce approximately equivalent blocks. Pupils set within blocks. This results in sets spanning a greater range of ability.
2   The year population divided into two, three or more ability bands and pupils set within the bands. This will usually result in sets covering a wider spread of ability. Low-attaining pupils in mathematics may be trapped in the lowest set of the next-to-lowest band where there is no provision for 'remedial' help.

Possible effects of setting on low attainers in mathematics
   a   Many of the effects of streaming apply also to setting. However some of the disadvantages are reduced, for instance:

no difficulty arises over pupils whose performance varies over different subjects, and hence there are likely to be fewer cases of individual injustice;
for the same reason setted groups are likely to be more homogeneous in mathematical ability;
movement between sets is likely to be easier since only one subject is likely to be involved;

the loss of self-esteem among lowest-set pupils is likely to be reduced if they are taught in mixed-ability groups for substantial portions of the week.

b   Particularly in primary schools, setting will create difficulties in following up the work and will tend to isolate mathematics from the rest of the curriculum because:

teachers will not be responsible for teaching mathematics to all the pupils in their class;
the timetable will be blocked, making a more rigid distinction between mathematics and other subjects.

c   Children, especially low attainers, may encounter difficulties because of the need to relate to different teachers and different pupils.

d   Because sets cover a limited range of ability, the class is often treated as a homogeneous group and always taught as a class despite the following limitations:

the selection techniques depend on subjective compromises so they cannot identify a homogeneous group;
pupils' attainment may vary in different aspects of mathematics; and
pupils' relative attainment varies over a period of time.

TEAM TEACHING AND/OR THE USE OF A FLOAT TEACHER

This is most frequently associated with open-plan primary schools or schools using mixed-ability grouping, but it is occasionally found in other schools.

Possible effects of team teaching on low attainers in mathematics
a   It permits flexibility in:

the size of teaching groups;
the selection of pupils for groups;
the amount of help given to low attainers; and
the work given to low attainers.

b   It makes great demands on teachers' time because of the need to discuss and decide policies relating to:

allocation of responsibilities in the classroom;
agreement of teaching approach and control of pupils – this organization is of limited value if a whole-class teaching approach is used; and
planning work and sharing observation of pupils.

c Teachers must be able to cooperate and work closely together.

d The teachers' skills may be developed as a result of working together and discussing and resolving common problems.

Possible effects of the use of a float teacher on low attainers in mathematics
a If the float teacher is only used for part of the mathematics time and responsible for particular pupils, problems arise concerning work setting/marking and support in the absence of the float teacher.

b As this form of organization uses additional staff, other classes will have a poorer pupil:teacher ratio for at least some of the time.

c Low attainers are less likely to feel isolated or rejected, and can be included in the mathematics going on in the class.

d Continuity is easier to maintain than it is when pupils are withdrawn.

e The system is more flexible and allows changes in the group of pupils who need help.

f If the float teacher helps pupils in general and not just the low attainers, then there is unlikely to be a loss of status associated with receiving extra help.

How can resources be deployed to minimize ill effects of the organization?
Staff in the school may feel that, essentially, the organization of the school or mathematics department meets most of the major needs and that with a few adjustments the needs of the low attainers in mathematics can also be met. One of the major requirements, whatever the overall organization, is for flexibility in dealing with low attainers.

Points worth considering when trying to ensure that satisfactory arrangements are made for low attainers are listed below for each major type of organization.

FOR SETTING AND STREAMING
1 Giving considerable attention to selection procedures to ensure that pupils are placed so as to receive the most appropriate teaching.
2 Arranging favourable pupil:teacher ratios in lowest set(s) or stream(s).
3 Frequently reviewing pupils' progress and allowing mobility between sets or streams so that no pupil is held back.
4 Ensuring that the curricula of different sets or streams are related so that transfer between them is possible.

In an 11–18 inner urban boys' comprehensive school, all pupils were set for mathematics. In general pupils were class-taught and followed the SMP numbered or lettered series of texts.[1] The lowest set, however, was treated very differently. It was about half the size of the others and pupils worked at their own rates and levels using a wide variety of apparatus including materials for practical work on capacity and games to reinforce simple number knowledge. Book work was based on the individualized use of the *Hey Mathematics*[2] scheme.

A problem arose because of the considerable difference in approach between this set and the others. If a pupil in the lowest set improved considerably and was moved up to another set he was, invariably, unable to maintain the good progress because of the difficulties in adjusting to a different and more formal method of teaching with less individual help. In some instances pupils had to return to the lowest set.

To overcome the problem, the mathematics department created a transitional set which had a slightly larger number of pupils, a more formal approach and less individualized and practical work than the lowest set.

5 Building up the morale of the lowest set or stream by using attractive materials, pleasant room, interested teacher, etc.

In an 8–12 middle school where third- and fourth-year pupils were set for mathematics, the deputy head, who was also responsible for mathematics throughout the school, took the third-year lowest set. This teacher was enthusiastic, and very interested in mathematics. Her room was attractively decorated .There was a pleasant atmosphere, the pupils seemed highly motivated, responsive and interested in their work and there was no evidence of a 'bottom-set mentality'.

6 Where pupils are streamed, blocking mathematics on the timetable so that pupils whose attainment in mathematics is out of line with their attainment in English can be taught in an appropriate group.

7 Not allowing pupils to be placed in a low set or stream simply because they are disruptive.

FOR MIXED-ABILITY GROUPING

1 Providing opportunities for teachers of similar groups of pupils to meet and discuss classroom organization and record keeping.

2 Building up a variety of resources, particularly: (i) making or acquiring a sufficient supply of materials which are suitable for low attainers; and (ii) acquiring materials and/or educating pupils so that some aspects of the work can be organized by the pupils themselves, thus allowing the teacher to teach and not simply to organize.

3 Having more teacher contact with low attainers, for example: (i) grouping pupils in the classroom by ability and making the low-attaining groups smaller but spending a similar amount of time with every group; (ii) having a float teacher to help low attainers; (iii) withdrawing low

attainers in small groups from mathematics, or at other times, as appropriate. Withdrawal groups can vary from timetabled groups to a flexible arrangement where pupils having difficulty during a lesson go for help to an extra teacher specifically allocated to this task.

In an 11–18 comprehensive school, pupils were taught mathematics in their mixed-ability classes using an individualized learning scheme. The mathematics department had considered various ways of arranging withdrawal, the aim of which was to support the normal work of the pupil either by approaching the work in a different way or by teaching the underlying skills and concepts.

One idea was chosen and tried. This entailed withdrawing up to six pupils per class in the first, second and third years for one of their five mathematics lessons each week. The pupils who were withdrawn were described as being less able, slow working, disruptive, or those who had performed badly on a test. Although the composition of the groups was meant to be flexible, in practice the same children were withdrawn for the lesson each week. The mathematics staff liked the system but wanted it to be more flexible to accommodate both those classes in which few pupils needed to be withdrawn and others which had five or six pupils needing extra support for most of their mathematics lessons.

The department decided to alter the organization for the following year so that the withdrawal group would be available to pupils for every mathematics lesson. This could be achieved with only minor changes in the staffing allocation because classes were blocked for mathematics.

POSSIBLE EFFECTS OF WITHDRAWAL GROUPS
In general these comments also apply to the use of a float teacher helping groups within the class.

There are several ways that the withdrawal groups can be organized. Pupils can be selected for a variety of different reasons. They may, for example, be:

generally low-attaining;
having difficulty with a particular topic;
frequently absent from school;
generally high-attaining and thus needing work which extends them; or
needing more practical work and discussion than can be provided in the ordinary classroom.

The group may consist of the same pupils each time, different pupils according to their needs or a mixture of both. The objectives for the group can vary but will usually relate to the criteria for selecting pupils. Children may be extracted just for mathematics or for a variety of subjects. The frequency, duration and size of the groups can vary: pupils may be withdrawn only from the mathematics lessons or at other times. Withdrawal may be associated with each of the major forms of organization – mixed-ability, streaming and setting.

These are the possible effects of withdrawing pupils:

1 Pupils can be taught in very small groups.
2 Work can be matched to individual needs.
3 Pupils can be given more individual help.
4 There can be a more relaxed atmosphere and more conversation; it is easier to organize practical work for a small group.
5 A group of pupils may be identified who are at roughly the same stage and who could benefit from doing similar work.
6 Pupils can continue to be involved in the mathematics going on in their own class.
7 Pupils may dislike being withdrawn because they miss a favourite subject or one which they feel is important.
8 It is difficult to return pupils who make good progress to their class full-time since they often do not maintain their improvement. This may be because: (i) they are dependent on the individualized help given in the withdrawal group; (ii) the teaching approach is very different; or (iii) they have missed some important work.
9 If low attainers who are slow learners are withdrawn it is unrealistic to expect that they can ever catch up.
10 The organization can make use of available staff expertise.
11 The class teacher may opt out of responsibility for the pupils.
12 Withdrawal groups can only be arranged by increasing overall class size or by reducing the number of non-teaching periods for staff.

Table 3 summarizes the different ways in which extra help can be given to low attainers.

It may be that the overall organization for the teaching of mathematics requires only slight adjustment to make the system flexible and well suited to the low attainers in mathematics. In some schools, however, the organization is such that even with adjustments it cannot provide adequately for low attainers in mathematics. If this is so, then it follows that the organization needs to be carefully revised. Often the single most effective action which might be taken for the low attainers in mathematics is to change the overall school organization. It is important to remember that organization is a tool, there to facilitate teaching and not to hinder it; that it is not fixed but can and should be changed when the need arises. On the other hand, imposing an organization on unwilling teachers is unlikely

Table 3 Summary of ways in which extra help can be organized. Usually the least able pupils are withdrawn, but there are many advantages associated with withdrawing other groups of pupils and leaving the class teacher with a small group which includes the low attainers

Extra help may be given:	Extra help may be given in work which is:	Pupils given extra help and choice of work may be determined by:	Extra help may be given:	Other variables include:
to the same pupils every time	same for all class	class teacher	in classroom	frequency of help
on an *ad hoc* basis of who needs help with particular task/topic	related to classwork (e.g. general topic)	float teacher	outside classroom	duration of help
	quite separate for group receiving extra help	headteacher	either, depending on activity	range of curriculum covered
		specialist teacher		size of group
	mixture	planned cooperatively between any of these		

to benefit the pupils. Rather, the need for a change in organization should arise from teachers' perceptions of how the pupils should be taught.

What effects does the classroom organization have?

Whatever the overall form of organization there will be a range of ability, rate of learning and attitude towards mathematics in every class. Similar forms of organization within the classroom are likely to affect low attainers in a similar way whatever overall school organization is adopted.

TIMETABLING

In secondary schools mathematics lessons are usually allocated to a particular time on the timetable but this need not be the case where pupils have the same teacher for many subjects.

In a first-year junior (7- to 8-year-old) classroom the teacher felt that pupils benefited from discussing their mathematics work with her so she usually organized the day's work so that only one group of about six to eight pupils were doing mathematics at any one time.

In a banded comprehensive school the remedial department was responsible for teaching all non-practical subjects to pupils in the bottom band. Because the same remedial teacher took a group for a number of subjects, the amount of time spent on mathematics was fairly flexible. In one case the teacher had organized the day so that following the morning break after a double period of mathematics, pupils were able either to continue with their maths topic work or to get on with their reading book.

Flexible timetabling is nearly always possible in the primary school for at least some of the time. Pupils may be able to study mathematics at almost any time in the school day and not all pupils need do mathematics at the same time. This flexibility could have many effects:

1　The teacher can spend time with a small group of low attainers in mathematics while other pupils get on with work which can be done without the teacher's help.
2　Less duplication of materials is needed if not all children study mathematics at the same time.
3　This type of organization lends itself to the study of mathematics across the curriculum, for example in environmental work or art and craft.
4　In such a flexible organization it would be easy for pupils to work at mathematics for varying amounts of time rather than always being restricted to the same length of lesson.

5 The teacher would find this considerably more difficult to organize in terms of:

controlling pupils;

keeping a record of what each child has done; and

making sure all pupils get an adequate amount of mathematics.

ORGANIZATION OF PUPILS

At a particular point in a mathematics lesson, each pupil is either being taught or discussing work with the teacher, or working independently of the teacher. In either case the pupil may be working as an individual, as a member of a small group, or as a member of the class. The proportion of time spent in any one of these six ways will vary considerably from one class to another. For example in formal class teaching each child will function as a member of the class when interacting with the teacher and as an individual when working without the teacher; in other classes pupils may work in each way on different occasions depending on the task.

Individual work

a Opportunities for expressing ideas mathematically, discussing work and cooperative planning will be lacking if pupils always work individually.

b Some deliberate provision must be made for encouraging mental work, e.g. immediate recall of number bonds, mental addition of two-digit numbers, etc. This could be done either by having some class activities or by using a tape recorder.

c Those pupils who are disruptive or have emotional difficulties often find it difficult to share and work with others. They are likely to work better on their own and may interact more positively with the teacher on an individual basis.

d The work can be tailored to suit each child, although, in practice, the extent to which this is possible will be limited by the teacher's time and knowledge of the pupils.

e The teacher may find that a class of children working individually makes so many administrative demands in terms of marking, interpreting the task, etc. that there is little time for teaching.

f The materials, especially those which are used by low-attaining pupils, will need to be chosen very carefully. In particular the level of reading difficulty must be matched to pupils' ability, and ways must be found to present work for non-readers, e.g. tape recordings.

g Teachers will need to organize some systematic checking procedure to ensure that all pupils maintain a reasonable pace of working, since it becomes more difficult to spot immediately those children who are working only intermittently on their tasks.

h More practical work may be possible since it is no longer necessary to have class sets of apparatus, calculators, etc.

Group work

There are many ways of grouping pupils. As well as variations in group size, children may be chosen for a particular group in a variety of ways, e.g. similar ability, widely differing abilities, friendship. In addition, where group work is undertaken, pupils may always work in the same group or they may work in a group whose size, method of selection, and so on depend on the task.

Essentially group work may be of two different kinds: pupils may be grouped together to work on a shared task, as, for example, in problem solving or investigations; or they may simply be grouped together when the teacher is with them for discussion, introduction of new work, or teaching a particular point, after which they return to working individually or in pairs. Sometimes children are put in groups and each group is set similar work yet there may be little or no interaction between them.

Below are some points which need to be considered in organizing group work for low attainers in mathematics.

a Low attainers are often less mature than other pupils and may not participate.

b Low attainers may take a passive role, especially if the group is mixed in ability.

c In a mixed-ability group, it is difficult to select tasks in which the low attainers can both participate and benefit.

d Working in a group permits discussion of work, sharing ideas, learning by observing other pupils' mistakes, etc. This may lead to more relational understanding as discussed in Chapter III.

e Group teaching allows the teacher to collect together pupils who are at a similar stage of learning and devote more time to direct teaching.

f Group work makes demands on the teacher in terms of organizing pupils so that attention can be given to one group rather than the whole class.

g Low attainers, in particular, are often more responsive in a small group: not overwhelmed by the teacher as may happen in a one-to-one situation, or by the rest of the class, particularly by the more able and extrovert pupils, in a formal class lesson.

h Low attainers may persevere longer if they have support from other pupils in a group.

Despite the difficulties which have to be overcome before low-attaining pupils can begin to work cooperatively with each other and with more successful pupils there are advantages for the pupils if group work is undertaken. The discussions which take place in planning and interpreting a task encourage children to think, express themselves and listen to others. Pupils benefit socially as well as intellectually from such activities.

Class work

a It is often inappropriate to class teach since, unless the range of ability is very narrow, what is taught will only be at the right level for a few pupils and almost invariably too difficult for the low attainers in mathematics. Many low attainers simply 'switch off' when addressed in this way, through long experience of 'not understanding'.

b If something the teacher wishes to teach is at an appropriate stage for all the pupils and is sufficiently motivating that many are likely to pay attention, then class teaching will save time. This is often the case when the teacher is trying to give an 'overview', for example the introduction of a new topic or a discussion of the social significance of some aspects of mathematics.

c Class instruction should always be brief, particularly for low attainers. Even adults have difficulty in concentrating for a long time.

d Working as a class can involve all pupils in presenting and describing work they have done, for example, class construction of a large item for display. However there is a danger that low attainers may be relegated to a passive or very insignificant role.

It is valuable that a proportion of the mathematics time should be spent in each of these ways – individual, group and class work. Inevitably a teacher finds that some forms of organization suit him or her more than others, but constant repetition of one preferred method will not provide adequate variety. Low attainers, in particular, need this variety, partly because change is itself stimulating, but primarily because to experience the different mathematical topics which form a broad curriculum, and to cover the different aspects of mathematical learning described in Chapter III, demands a wide variety of activities which cannot all be provided within one form of organization.

ORGANIZATION OF THE CLASSROOM
The classroom needs to be organized to accommodate the planned work.

If the pupils work at different times as individuals, members of small groups or members of the whole class, then it may be necessary to rearrange seating and tables to facilitate this. Apparatus, discussed more fully in Chapter IV, must be stored so that it is easily accessible. Allocation of space must be considered in relation to the practical activities, the discussions and the quiet individual work which is planned.

ORGANIZATION OF THE TEACHER'S TIME

In Chapter III it was argued that pupils need:

> work at an appropriate level;
> material-based learning;
> questioning and discussion;
> problem solving and open-ended questions as well as the other aspects of mathematical learning; and
> a wide range of mathematical ideas.

These points give some idea of what pupils will be doing in the classroom; as such they have implications for the way in which the teacher plans lessons and uses the teaching time. As well as demonstrating and explaining there are many other aspects which influence the effectiveness of the teaching.

To arrange that pupils have work at an appropriate level the teacher will need to assess pupils by observation, testing or both. This is considered in more detail in Chapter VI. It will involve observing and talking to pupils.

Encouraging children to be responsible for finding their own materials and replacing them afterwards will reduce the amount of time which the teacher spends on organization within the lesson, but it will make demands on the teacher in pre-planning, such as labelling materials and organizing storage. The teacher's time may also be used more beneficially if pupils are given tasks which carry over to later lessons, and activities are planned so that not all the pupils start a new piece of work at the same time. Similarly if the teacher wishes to spend time with an individual or small group then the tasks which other pupils are given should be ones that can realistically be tackled without teacher assistance. Activities such as worksheets, puzzles and games, or work connected with a long-term project could be made available so that pupils need not continually interrupt or waste time waiting for the teacher. Sometimes a pupil who has had some success with practical work, a game or a puzzle can introduce it to another child, not only allowing the teacher to do something else, but also reinforcing the pupil's own knowledge and self-confidence.

Whilst it is vital that work done by a pupil be checked and that the pupil know that this is done, it is desirable that marking be kept to a productive minimum. Some activities can be largely self-checking; on other occasions pupils may mark their own work and this should help to encourage pupils' responsibility for their own progress.

In any attempt which a teacher makes to improve the effectiveness of his or her teaching, it is worthwhile to observe the pupils to see how they spend their time and to assess whether this matches what the teacher thinks they should be doing. Similarly it may be valuable to step back and analyse the time spent on teaching activities both in the lesson and in lesson preparation, and to see if this reflects the teacher's beliefs about how his or her own time should be spent.

Points for discussion

1   What do you consider are the main drawbacks of the school organization for teaching mathematics to low attainers? Is there any way of compensating for these?
2   Is the system flexible enough to cater for the range of needs of individual low attainers, e.g. non-readers, frequent absentees, slow learners?
3   Is there special provision for low attainers? If so are the pupils involved reviewed regularly? Can they transfer to the mainstream where appropriate? Can others, who fall behind later, still be helped?
4   Is the school, department (at secondary level) and classroom organization reconsidered regularly, both as a matter of course and in the light of particular changes in circumstances?
5   What do you consider are the main disadvantages for low attainers in mathematics of your chosen form(s) of classroom organization? Is it possible to vary this organization or compensate in other ways?
6   What proportion of the teacher's time in the classroom is spent in:

teaching;
marking;
administrative work;
controlling pupils; and
observing pupils?

Is this the correct balance?

References

1. School Mathematics Project, *School Mathematics Project: The Main School Course* (Books A–H). Cambridge University Press, 1968–72. Also *School Mathematics Project: Books 1–5*. Cambridge University Press, 1965–69.
2. L. Svensson *et al.*, *Hey Mathematics* series (Modules 1–6), 1974–75. English edition, J. Boucher, General Editor. Holt, Rinehart and Winston.

VI. Assessment and recording

Why assess low attainers in mathematics? What methods are appropriate?

Pupils are assessed to determine their 'present state of knowledge'. Different types of assessment were used by schools; some, notably in the primary sector, used only informal observational methods whilst others used combinations of written tests, oral tests and observational techniques.

The value of assessment lies in the use which is made of the information it reveals. It is therefore of paramount importance to decide why the assessment is to be made and then to consider, in the light of this information, what adjustments should be made to the organizations, courses or teaching approaches. Acquiring information without the possibility of action would be time-wasting for both teachers and pupils.

The major uses of assessment include:

to identify pupils in need of extra help;
to allocate pupils to classes or groups, and sometimes to indicate what forms of organization may be appropriate;
to help the teacher plan appropriate work, either by selecting particular activities or by allocating pupils to particular courses;
to evaluate the effectiveness of the teaching;
to aid continuity when pupils transfer to a new teacher or school;
to provide information for parents and prospective employers;
to motivate pupils;
as a way of controlling the curriculum; and
to compare school standards from year to year and with those of other schools.

The school described below had designed different forms of assessment to serve different purposes as part of the policy for low attainers in mathematics.

One 11–18 comprehensive school had a compensatory department for advising and assisting with the teaching of low-achieving pupils throughout the school, as well as teaching pupils who had special problems with language and/or mathematics.

In order to select the children who were to receive remedial help in mathematics, the school considered the results of an arithmetic test which had been given to all the incoming pupils in the fourth year of their junior school, the reports received from the junior school, and the results of a screening test which was given to pupils in the first two weeks at the secondary school. The latter was an oral test so that the effects of poor reading ability were minimized. It contained sections on 'number', 'money', 'time' and 'measure'. The test had forty questions. The staff would have liked to give remedial teaching to all the pupils who scored less than thirty, but in most years the number of pupils who could be accommodated was the determining factor. Each year, the remedial department accepted responsibility for the bottom thirty pupils, so the 'cut-off point' in the test varied from year to year.

The pupils who were selected for full-time remedial help in mathematics were then given diagnostic tests in 'number', 'money', 'time' and 'measure'. These tests, presented individually with the teacher reading out the questions, attempted to assess concepts as well as facts and skills.

There was, for instance, a question to see whether each pupil could conserve number. The teacher was asked to say to a pupil:

'Here are two rows of pennies.'

'There are the same number of pennies in each row.' (Indicate this by matching one to one on the magnet board, and then spread out the bottom row of pennies.)

'Are there now *more* pennies in the bottom row than the top row?'

'Are there now the *same number* of pennies in the bottom row as in the top row?'

'Are there now *less* pennies in the bottom row than the top row?'

'Put a ring round the right answer.'

In the measure test, pupils were shown a picture of four glasses of lemonade and asked: 'Here are some glasses of lemonade. Put a ring round the glass which is a quarter full.'

Near the end of the time test children were asked: 'This is 1977. Today is Tom's twelfth birthday. In what year was he born?'

The results of these tests were used to plan individual programmes of work for each child.

The aim of the compensatory department was to get pupils back into the mainstream as soon as possible. Teachers' observations and considered comments were used to make decisions, usually at the end of each academic year, about who should transfer.

Each of the reasons for assessing pupils listed above will now be discussed in more detail and illustrated with examples from schools. Many of the examples involve more than one method of assessment, and the use of the results is often not confined to that suggested by the heading which it is chosen to illustrate. Perhaps inevitably, many of the examples relate to the more formal types of assessment since it was easy to gather this type of information in a brief visit. However it should be remembered that most assessment is informal, and indeed is often at the immediate level of being used by the teacher to judge what question to ask next.

TO IDENTIFY PUPILS IN NEED OF EXTRA HELP

In a first school all the pupils did the Leicester Number Test[1] at the beginning of their last year. Pupils who scored badly on this test were given some extra help in mathematics in a withdrawal group.

In an 11–18 secondary school which used SMILE,[2] primary records of the new first-year intake were examined and each pupil was then set an initial group of tasks which teachers felt would be appropriate. Teachers observed pupils very carefully at the beginning, noticing how easily they carried out the tasks, supporting pupils if they got into difficulty, if necessary by providing alternative tasks. In addition teachers looked out for pupils who had problems in coping with the organizational demands made by an individualized learning system. Those who were unable to cope were subsequently taught in small groups by a remedial teacher.

The term 'screening' in the sense of sifting is often used to describe the process of picking out pupils who are at risk. Pupils who are not identified are those whom teachers expect to benefit from the normal provision while those identified by the screening will be assessed in greater detail. According to the complexity and flexibility of the organization several different stages of assessment may be necessary to arrive at a final identification of the group of pupils who are to be given extra or different help. Where the organization is very flexible, as in many primary schools, the teacher may initially identify those pupils who obviously are weak and make particular provision for them, while still observing the rest of the pupils carefully in case children who have learning difficulties have been missed. In the light of these further observations a reassessment can be made and appropriate adjustments carried out to the organization and planning of work. In a more formalized organization such as that in a secondary school which has a fixed-size remedial class, it may be best to do an initial rough screening right across the year group which will pick out quite a number of pupils each of whom can be given a second, more detailed assessment in order finally to select those who would benefit most from being in the remedial group. This is illustrated by the following example.

In a 13–18 comprehensive school, the remedial department had written guidelines for assessing pupils who had been initially selected for one of two parallel remedial classes on the basis of middle school reports. During the first four weeks of term each class spent their mathematics lessons for two weeks on formal computation, and for the remaining two weeks they studied the topic 'family size'. The remedial teacher was asked to assess pupils' abilities which emerged in studying the topic, doing homework sheets and mental arithmetic, particularly with respect to:

> 'lack of number concept;
> poor basic skill in $+$, $-$, $\times$, $\div$;
> lack of ability to see any relationship between tasks;
> poor physical control;
> lack of ability to follow work through to a conclusion;
> poor memory;
> little mathematical experience;
> lack of interest and enjoyment in the subject;
> homework refusals;
> language problems, particularly for immigrant pupils;
> good grasp of concepts and completion of tasks but poor organization of work and written expression.'

This information was used to group the pupils more appropriately, plan further work and, sometimes, to recommend that a pupil should move into a CSE stream.

Identification may be done by testing, observation or a mixture of the two. For an initial screening a *'norm-referenced' test* may be sufficient. This type of test is designed essentially to produce an ordering of pupils; each pupil is assigned a mark which will allow a comparison with other pupils but the mark itself will not convey very much. Most general tests set by teachers, such as an end-of-year examination, belong to this category, as do a number of published tests. Some of these published tests cover a broad spectrum of mathematics topics and include both understanding of concepts and performance of skills (e.g. NFER *Basic Mathematics Tests*[3]); others are restricted to number (e.g. the *Leicester Number Test*[4]), and some include only formal computation exercises (e.g. SRA diagnostic arithmetic tests[5]). Some children who perform adequately on computation tests but have insufficient understanding to be able to apply their skill are therefore more in need of remedial help than other pupils, who may have a good basis of understanding but have temporarily forgotten the procedures for, say, long multiplication and long division. Therefore if a norm-referenced type test, whether commercial or home produced, is used for the initial screening, the broader its content and approach the better.

Many published tests are also standardized, that is, the results have been analysed on a large sample of children and hence it is possible to say how a child with a given score compares with the total child population of that age in England and Wales. Thus such a test can be used for picking out all children who would be in the lowest 20 per cent on a national scale, at least with reference to their scores on that particular test.

Some problems arise with norm-referenced tests. It is difficult to infer any real meaning from a particular score since the same mark may have been achieved in many different ways. Also the rank order of pupils may vary quite dramatically from one norm-referenced test to another; the scores reflect the balance of the test in terms of both its mathematical content and the type of mathematical abilities tested.

This again suggests that if such a test is used for rough screening purposes the group of children initially identified should be substantially larger than the group for whom extra help, if any, can eventually be provided. To make decisions about exactly which children need extra help, and what sort of help they need, teachers may wish to use a test which is designed to determine pupils' levels of attainment in particular skills or concepts. Such tests are known as *'criterion-referenced'* tests. They may cover a fairly wide area of mathematics, for example the NFER *Essential Mathematics* test[6] which is administered individually to pupils. It aims to identify fundamental gaps in the secondary pupil's knowledge and the accompanying notes suggest appropriate remedial action. It asks the pupil, for example, to decide which of two numbers is the larger.

Other criterion-referenced tests cover key mathematical areas in more detail, for example the ILEA *Checkpoints*.[7] Many of the Checkpoints need to be administered individually; they are suitable for children of primary age and for low-attaining secondary children. Each Checkpoint deals with one basic concept or idea. For example, understanding and use of the way we write numbers is tested partly by seeing whether a child can immediately write down the number of cubes on a table when they are arranged in groups of tens with some loose ones, or whether he or she has to count all the cubes individually.

The NFER *CSMS Diagnostic Maths Tests*[8] are also criterion-referenced. Each aims to test understanding of concepts in a broad area of mathematics, e.g. measurement, place-value and decimals, reflection and rotation, number operations, graphs, etc. They are suitable for secondary school children and can be administered either to a class or individually.

Since most of the criterion-referenced tests referred to take longer to administer and analyse than a norm-referenced test, it may be appropriate to use them only with a group previously identified by a screening procedure.

Observational procedures for identifying pupils at risk may be formalized, such as the classroom observation procedure (COP)[9] developed by the Inner London Education Authority (ILEA) for use with 6- to 8-year-olds. Experienced teachers may be able to use informal observation procedures to pick out pupils who are having difficulty and determine the nature of these difficulties by, for example, seeing whether a pupil:

can count accurately;
if briefly shown a collection of twelve conkers and asked to estimate the number, gives a sensible reply and, if two are taken away, changes the estimate appropriately;
knowing that $5 + 8 = 13$, can use it to find $8 + 5$,
$$13 - 5,$$
$$5 + 28;$$

can 'read' 2-digit and 3-digit numbers.

TO ALLOCATE PUPILS TO CLASSES OR GROUPS

In most schools visited, both primary and secondary, pupils were grouped according to ability for mathematics either in sets or in groups within the class.

In a large junior school, there was a remedial stream in each of the four years. Pupils were selected for the first-year remedial class chiefly because of poor reading; this was done in consultation with the feeder infant school. First-year teachers from the junior school visited the infant school and discussed which pupils the infant teachers thought would need extra help in reading. Staff in the junior school used this information to select pupils for the remedial class.

In the following school more than one method of assessment was used to group pupils:

On entry to an 11–18 boys' comprehensive school, primary records were used to allocate pupils to one of three ability bands. Soon after the beginning of term all first-year pupils were given a test which had been written by the remedial department. This was used as a check on placement. It was also used for pupils in band three, the remedial band, to identify those who were having especial difficulty and the particular ideas which they had not fully grasped.

The test was designed to assess facts, skills and concepts in different aspects of mathematics and at different levels and included, for example:

'Which jar has the most jam? Put a line underneath it.'

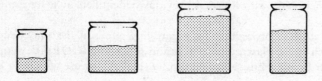

'Write today's date.'
'When is the next leap year?'
'How many days make a leap year?'
'Draw a line 10 cm. long.'

In a junior school, pupils in one of the second-year classes worked in friendship groups on a topic on transport which was developed across the curriculum in the first few weeks of the school year. By the time that most of the planned work had been completed the class teacher had gained an initial insight into her pupils' levels of attainment in different aspects of the curriculum which enabled her to allocate pupils to groups for English and mathematics according to their ability in those subjects.

After consultation with the feeder first school, *the staff of a middle school* decided, prior to the beginning of each year, how they were going to organize pupils in the incoming first year. The decision was made on the basis of how many pupils were recommended as needing remedial help in English, reading and mathematics. The teachers were able to consider various arrangements, including a permanent remedial class, withdrawal groups for one or several subjects, and a float teacher in the classroom.

Many of the tests described above can be helpful in providing a source of information which teachers may wish to use in grouping pupils. It is important that any test chosen match the teaching aims; thus it may be better for the school to draw up its own test rather than use an inappropriate published one. In particular, the range of the test in terms of both mathematical content and type of mathematical activity should be as wide as possible so as to be fair to pupils with different strengths, especially if they come from a variety of backgrounds.

The points made above also apply here; it should again be stressed that a decision based only on the results of a norm-referenced test is likely to give rise to very arbitrary placements. For example two children may be put into different sets and thus follow essentially different mathematics curricula, on the basis of a difference in score of one or two marks, when, on a different test, the child with the lower mark might have scored much higher than the other child.

For this reason, and because pupils progress at different rates, the allocation of pupils to groups needs to be reviewed regularly on the basis of further assessments, especially teachers' observation.

TO HELP THE TEACHER PLAN APPROPRIATE WORK

In a three-form entry girls' secondary modern school, pupils were allocated, alphabetically, to mixed-ability classes. KMP[10] was used in most of the mathematics lessons. On entry all the first-year pupils were given a test in which each question was related to tasks in the KMP programme. The results of this test indicated which tasks would be suitable for each pupil and were used to set their initial work.

In the top infant class of a JMI, Brian was working from Fletcher's *Mathematics for Schools*[11] Level 1 and had been asked to partition a set of seven objects in different ways and complete number sentences such as $2 + \square \rightarrow 7$. At the teacher's suggestion he counted seven crayons and was asked to split them into two groups. Brian divided the crayons into a group of three and a group of four. He counted the group of three and said, 'Three there', and then counted the group of four: 'Four there'. When asked how many altogether he put them together again and counted from one: 'There are seven'.

The teacher asked him if he could split them up in a different way and he made a group of two and a group of five and found the number in each group by counting. This time when asked, 'How many altogether?' he put them together, started counting from one again, but miscounted and said, 'There are six'.

As a result of this conversation, Brian's teacher decided that he should not be left on his own to do the formal written work which the textbook suggested. Instead she gave him a set of cards on which the numerals nought to seven were written and asked him to go on splitting the seven crayons in different ways but to use the cards to show her what he had done.

In many schools the progression of work within a scheme seems to be followed rigidly as if on the assumption that pupils who have done the work have necessarily understood it. Yet during the visits to schools, the team observed many children doing work which was meaningless to them because they had not acquired earlier concepts and skills which would help them to make sense of it. Providing appropriate material for a child is a matter of continuous adjustment.

Occasionally, especially when meeting a new set of pupils at the beginning of a school year, or when moving to a new topic, the teacher may feel the need for a global assessment of what a child can do over a relatively broad mathematical area. For instance, a teacher may present a child with a series of brief assignments on many different aspects of length and its measurement. These may range from whether the pupil recognizes that two pencils which are shown to be the same length when put together are still the same length even if one of them is moved along or turned round; to the estimation of the distance between two local towns in kilometres; or even to accurately measuring a cupboard giving the answer both in metres (using decimals of a metre) and in millimetres. The results will act as a very rough guide to what the pupil has mastered in terms of concepts and skills, and can be used to indicate the broad areas in which his or her knowledge may most profitably be extended by the provision of suitable experiences. Criterion-referenced tests of a broad kind, for instance the NFER *Essential Mathematics* test[12] or the NFER CSMS tests[13] already referred to on page 117, may be helpful here but there is not much assessment material of this kind available and groups of teachers may need to construct their own.

Teachers will need to 'home in' on particular topics in rather greater detail, possibly as the result of a more global assessment, in order to determine more accurately what work is appropriate for a particular child. This may be done using more specific assessment material such as the ILEA *Checkpoints*[14] or the teacher may decide to try the pupil on a particular assignment and observe the extent to which he or she is able to cope with it, using questions to determine whether the level is about right, or whether the child needs to jump forward several steps or return to much more basic ideas.

Even when the teacher is fairly confident of having found an activity which is just at the appropriate level, the pupil may not appear to make any progress. It will then be necessary to assess quickly, by questioning, whether the cause is relatively trivial – e.g. misreading the instructions – or whether the child is being held up by some fundamental mathematical misconception which is crucial to the activity. This type of assessment is *diagnostic* in nature. Unfortunately the term diagnostic is often used in a narrow sense to refer to number skills. For example, the teacher may diagnose that a child who does:

$$
\begin{array}{r}
34 \\
\times\ 20 \\
\hline
680 \\
34 \\
\hline
714
\end{array}
$$

is getting wrong answers to long multiplication because of not being able to multiply by zero correctly. However at a deeper level it might be said that the child has little understanding of place value and, in particular, of the effect of multiplication by ten. With such understanding the pupil should be able to arrive at the answer mentally. The two alternative diagnoses would lead to very different prescriptions in terms of follow-up activity. The latter might, for instance, suggest making enlargements of 'buildings' constructed out of centimetre blocks on a scale of 20 : 1, whereas the former would just lead to 'more sums'.

The whole area of assessment and diagnosis assumes that the teacher has access to a hierarchical structure of concepts and skills which are required for mastering any mathematical topic. This may be something which an experienced teacher will know, or he or she may decide to refer to a written document such as the school mathematics scheme, a set of LEA guidelines, or the published framework of a curriculum scheme.

Unfortunately, whereas a teacher of remedial reading is now likely to have a range of specialist diagnostic materials available to assess the specific needs of poor readers, there seems to be very little available in mathematics except that which has already been mentioned.

It is important that the relationship between any such assessment material and one or more teaching schemes be explicit. Links may be indicated by those producing the material or by established groups of teachers wanting to use it. This will help to ensure that assessment will lead to more appropriate learning experiences.

Some teachers in the schools visited were aware that assessment was one of the areas in which their teaching was weakest, and that as a result they had little knowledge of which mathematical ideas were understood by individual low attainers in their classes, and which fundamental ideas were lacking. This restricted the extent to which teachers could help their pupils and made lesson preparation a very 'hit-and-miss' affair, leading to gross mismatches between the children's capabilities and the curriculum being offered to them. For instance children may, like Hayley (Chapter III), be struggling with long multiplication when they actually know very few number bonds, or may be introduced to simple trigonometry when they are still at the stage of confusing 'ten times as long' with 'ten centimetres longer'.

Individual teachers may wish to reject the syllabus or textbook in favour of returning to more fundamental ideas backed up with practical work, and results of diagnostic assessment may enable them to convince other teachers that this is the best course of action for some children.

Too often teachers assess pupils only on those aspects of learning which are easily tested, namely facts and skills. This may have a restrictive effect on the curriculum, since the acquisition of facts and skills may become regarded as the most important aspect of mathematical education. Understanding of concepts may be assessed by the way that the pupil answers questions, using items, for example, from the CSMS tests,[15] *Checking Up*[16] or *Checkpoints*.[17] Or the teacher may devise his or her own way of testing concepts, informally with a small group or an individual, or more formally, with the whole class at once.

To test primary pupils' early number concepts, a teacher set out, on the floor, three collections of objects: thirteen large bricks, fairly well spaced out, twelve marbles, very close together, and fourteen cylinders of a size in between the marbles and the bricks. Pupils were asked to say which collection contained the most objects.

Some children responded intuitively, either to the large area covered by the bricks or to the density of the marbles, or guessed the cylinders because of their 'in-between' qualities. But only the pupils who made a reasoned comparison, by counting or matching one-to-one, indicated a grasp of the fundamental concept of number.

Since applications, investigations and problem-solving often do not lead to a single correct solution and may even be open-ended, their assessment is complex, often involving observation or discussion. But they will give the pupil an opportunity for reinforcing the facts, skills and concepts which he or she is supposed to know as well as giving the teacher a chance to assess the pupil's overall knowledge and ability to relate one idea to another.

Assessing pupils' attitudes

Before pupils can learn, they have to become actively involved, and in order to apply their knowledge they need to feel that the mathematics which they know is useful and can yield solutions. Since pupils' feelings about mathematics are a major factor influencing their ability to learn, the school needs to foster positive attitudes. Low attainers may be anxious and lacking in confidence; overtly they are often aggressive or apathetic. An assessment of both the negative attitudes and the hopeful and positive ones could help the teacher to build up the pupils' confidence and self-esteem. Therefore discussion with individual pupils about how they feel about what they are learning and when appropriate, particularly at secondary level, asking pupils to write or talk about this, is worthwhile. In addition the teacher may determine the pupils' particular anxieties by noting their reluctance to attempt specific tasks.

TO EVALUATE THE EFFECTIVENESS OF THE TEACHING

When appropriate work can be planned on an individual basis the means of assessing the pupil may also be used to evaluate the teaching approach since ultimately it may be neither the pupil nor the teaching approach which is 'wrong', but only the suitability of one for the other. Since the pupil remains, it is the teaching approach which has to be changed, a fact which can be very hard for teachers to accept especially when, with the older pupil, it implies that very elementary work is required.

Assessment may be used throughout the school to evaluate teaching approaches, but this is usually done over a fairly long period of time – a year or maybe a term. Since success is important to the progress of low attainers, it is preferable that the teaching which they receive be evaluated on a shorter time scale, even to the point of immediate feedback – 'quality control'.

In one coeducational 11–18 comprehensive school where all the mathematics teaching groups in years one, two and three covered the full ability range, each pupil's allocated work usually consisted of ten independent tasks. It was felt to be very important that pupils be seen on an individual basis to discuss their work with a teacher, so that pupils who had been set inappropriate work and were in great difficulty could be given a modified or different task. In addition, after completing their tasks, each pupil did a test in which questions were related to the original tasks. In marking the test, the teacher assessed the pupil's attainment in each aspect of the work and the new tasks were set in the light of this assessment: each could be at an easier, similar, or more advanced level than the previous ones.

TO AID CONTINUITY WHEN PUPILS TRANSFER TO A NEW TEACHER OR SCHOOL

Although some teachers reject the judgement of others (a point discussed in Chapter I), if significant details can be communicated to the new teachers, valuable time can be saved and pupils may be spared inappropriate experience. Such information may also be helpful in assessing the pupil's potential. However, the effectiveness depends to a considerable extent on the method of communication. The way that information is recorded is considered in greater detail later in this chapter.

TO PROVIDE INFORMATION FOR PARENTS AND PROSPECTIVE EMPLOYERS

Public examinations, realistic school reports, profiles and leaving certificates, as well as indicating pupils' levels of achievement to prospective employers, can be helpful in guiding pupils and their parents to investigate suitable jobs.

In one 13–18 coeducational comprehensive school, pupils in the first three years were given tests in basic numeracy. It was suggested that some of the topics should not be attempted by the less able pupils; some were also considered unsuitable for the majority of pupils in their first year in the school. There were forty short tests, grouped under the headings Whole Numbers, Decimals, Fractions, Fractions and Decimals, Money, Measures, Percentages, Approximation and Ratio. These were given once in each year and pupils who succeeded in scoring nine or ten, the maximum in any test, did not have to repeat it in a subsequent year. The final mark awarded at the end of the three years was the best mark gained by a pupil in each test; all of these results were included in pupils' reports which were given to them on leaving school.

This school was banded, with two parallel bottom sets in the bottom band in each of the three years. Pupils in bottom sets in years two and three were not entered for a public examination. Instead, they followed the school's leaving certificate course, which was chiefly concerned with arithmetic and social mathematics. The assessment of the certificate was in three parts. Part 1 consisted of the results of basic arithmetic skills tests; these tests were in addition to those set in the mainstream. Pupils were given explanations, then worked through examples. After an interval of about a week they attempted the appropriate test. This was done so that the test did not simply assess immediate recall. The pass mark was 80 per cent and those who failed did more work and tried the test again later.

Part 2 was the result of a one and a half hour written examination in social arithmetic which was taken in the second term of the third year and reflected work which was covered in the second and third years.

Part 3 involved the assessment of project work which pupils undertook in the first term of the third year. Each pupil chose five projects from a list of twelve and received help in deciding what could be included in the selected area of work. A short report had to be written on each project selected.

Some of the schools visited believed that formal testing across a wide range of ability was of limited value in monitoring pupils' progress. Many of these schools, whilst not totally rejecting formal written tests, felt that continuous assessment was more useful as pupils were able to demonstrate their achievement over a period of time in types of work for which they might have greater aptitude, some of which could not be easily included in a written test. Whilst this argument refers to all pupils, it is especially relevant to low attainers who may find it very difficult to recall many disconnected bits of knowledge out of context in a limited period of time, as is required in most written examinations.

TO MOTIVATE PUPILS

Wishing to perform well in end-of-year tests or public examinations provides motivation for both pupils and teachers, but tests are usually more effective in motivating able pupils who perceive a likelihood of success. If this aspect of the work is heavily stressed, the low attainer may think that he or she will not succeed and give up completely. However, whilst written tests are unlikely to motivate low attainers, less formal methods of assessment, including observation and particularly discussion, often motivate pupils because of the personal attention which the pupil receives. Older pupils may also be motivated by project work which is to be presented for assessment.

AS A WAY OF CONTROLLING THE CURRICULUM

Once the objectives of the curriculum have been decided it may be possible to define them in terms of what the pupils should be able to do at the end of the course. However this will only help teachers and pupils if the objectives and the questions which pupils are to be asked are reasonable. Setting items which are too difficult pressurizes teachers into teaching instrumentally, while if items are too easy valuable teaching time is spent in unnecessary practice. It should also be checked that the assessment accurately reflects *all* the agreed curriculum objectives.

TO COMPARE SCHOOL STANDARDS

An individual school's concern about standards and comparisons with other schools is unlikely directly to benefit low attainers.

The tests which are likely to be used to compare standards in general assess mainly the acquisition of facts and skills and, as mentioned above, a demand for higher standards in these is unlikely to lead to more effective teaching for low attainers. On the other hand, an attempt to examine objectively the range of performance of pupils, such as that carried out by the Assessment of Performance Unit (APU),[18] may lead to a more realistic view of children's capabilities and, in particular, an appreciation of the extreme difficulty that many children have in learning mathematics.

What are the major problems of assessment?

Some of the schools visited, notably in the primary sector, found the idea of testing pupils quite unacceptable. This aversion to testing stems from a misuse of assessment in which tests are used to compare and subsequently to 'label' pupils in an unconstructive way which often becomes all too permanent, rather than being used to inform teachers about the pupil's stage of development which could be used to improve the teaching. Although there is a clear need for several different types of assessment, there are potential pitfalls in assessing pupils and their work. These dangers seem more apparent in written tests than in assessment by observation and discussion.

TOO MUCH IMPORTANCE MAY BE GIVEN TO TESTING

If testing is given too high a priority in a school, the pupil's learning may suffer in two distinct ways. First, too much time may be spent setting, administering, doing, marking and worrying about the tests, leaving less time for teaching and learning. Second, as noted above, because facts and skills are easier to assess than other aspects of mathematics, the school may over-emphasize these aspects and thus provide a restricted mathematics curriculum.

INSUFFICIENT ATTENTION MAY BE PAID TO THE LIMITATIONS OF THE TEST

As well as reading, carefully, the instructions about how a test should be administered, the age range for which it is suitable and other accompanying information, additional thought needs to be given to what actually is being tested so that no incorrect assumptions are made from the results. It is difficult to decide from a written test whether a fact, skill or concept has been tested – one child's rote knowledge is another's reasoned reply. On the whole written tests give a minimum of information about the pupil since an answer is usually either right or wrong and if it is wrong there may be little indication of why. Equally, if a pupil correctly works out $60 - 21 = \_\_$, does it mean that he or she can:

do all two-figure subtraction sums;
do this sum presented in a different way;
apply subtraction in real life; or
apply subtraction in written or verbal problems?

And if the pupil gets it wrong, or does not attempt it, what inferences can be drawn?

Sometimes tests are used simply to support the organization. For example:

In one comprehensive school, twelve places were allocated in the remedial department each successive year to the twelve first-year pupils who had the lowest scores in the *Richmond Tests of Basic Skills.*[19] This was not necessarily a true representation of their mathematical ability; since it tested only arithmetic facts and skills, it could not predict their potential with any certainty. Further, since the bottom twelve were selected, irrespective of an absolute score in the test, in some years fairly able pupils may have been needlessly labelled 'remedial' whilst in other years pupils who were very much in need of extra help may not have been selected if their scores did not lie among the lowest twelve.

Some difficulties can be avoided by matching the assessment to the required use. In general norm-referenced tests will be suitable for comparing school standards from year to year or with other schools. They will be of some value in identifying pupils in need of extra help, although this could be done in other ways with less damage to the self-esteem of pupils. They can be used to some extent to group pupils for teaching purposes but they will not identify a homogeneous group of pupils with identical or even similar needs.

Criterion-referenced tests, particularly those which test concepts, will yield more information about the stages of pupils' development, but in some cases little is known about the implications of particular concepts.

For example, a child who has not yet grasped conservation of number is clearly not ready to benefit from learning the number bonds by heart, but what this lack of understanding implies for providing teaching experiences is less obvious. In addition a concept is often partially understood: a pupil may demonstrate understanding in one context and not in another or may seem to have grasped the idea one day and forgotten it the next.

PUPILS MAY BE GIVEN TESTS THE RESULTS OF WHICH GIVE NO NEW INFORMATION OR ARE NOT ACTED ON

Pupils are usually worried about tests. The low attainer will often underachieve in a test, leading to a further increase in his or her lack of confidence and self-esteem. In many schools pupils who score badly in end-of-year tests still move on to the next year's course along with the pupils who have a confident and secure understanding of the work already covered. There can be little justification in giving a test if the results cannot influence the subsequent curriculum followed by the pupil.

INFORMAL ASSESSMENT MAY INCORPORATE SOME ELEMENT OF UNWITTING BIAS

The mathematical ability of articulate children is often over-estimated, as is that of children who work neatly and conscientiously. It is therefore a good idea to organize an occasional element of formal assessment to provide more objective information.

Recording

A written record of each pupil's progress is essential to aid continuity when the pupil moves on to a new teacher or a new school but also for the pupil's own teacher, since the information which is gathered about a pupil is often too much to remember.

The project team saw many examples of schools who received records from their feeder schools which were unused. Sometimes this was because the records were so detailed that it would have taken a very long time to read and act upon them; on other occasions they were so sparse that they had little meaning. In some of these cases the bases for the judgement were unclear; brief comments appeared often to be subjective and possibly biased by such other factors as those referred to above.

In one junior school, staff were not aware of how mathematics was taught in the infant school. LEA record cards sent on from the infant school described most pupils as having 'done: addition, multiplication, subtraction, division, tens and units, measurement of lengths, capacity and weight' but there was no indication of the level at which these had been studied, the way that the pupils had been taught, or the specific practical and written activities that had been undertaken.

For effective use to be made of such records, it is important that the content and format of the records should be agreed by the teachers who make the records and the teachers who receive them. This is also true of the school–employment interface; a committee on which are represented many of the major local employers, as well as teachers from one or more schools, is likely to provide a worthwhile forum for the design of records for school-leavers.

Experiences, attainment or a mixture of the two may be recorded. Noting pupils' experiences will depend on the teaching approach which is used so that a school which uses one standard textbook may simply record which pages have been successfully completed, but a more complicated method will be required by a school which uses a variety of resources.

Potentially a record of attainment is of much more value than the record of experience, both in planning suitable work and in evaluating teaching approaches, but it is also much more difficult to do:

In one junior school, second-year teachers had devised a tick sheet for their pupils, and filled it in to keep a check on progress and aid continuity when the pupils moved on to a new teacher in the third year. After a year's use, they were abandoned. The teachers had been trying to record pupils' achievement rather than experience and the problems of attempting this seemed insuperable. Different teachers meant different things when they said that a pupil had grasped an idea. Children seemed to be inconsistent: one day it appeared that they could do, or understand, another day they seemed to have forgotten again. The teachers decided that their time was better spent in teaching and preparing work, and abandoned the tick sheets.

Another school managed to communicate some of each pupil's attainments as well as experience:

In a junior school which used a popular standard text, the teachers filled in a section of a record sheet for each section of the textbook when it was finished. This record listed the pages which had been successfully completed and noted any difficulties which pupils had experienced. Notes were also made in a different column on the same sheet about any other related work which the pupil had done, from another textbook, from a worksheet made by the teacher, or as part of a general topic, developed across the curriculum. In addition, a folder was kept for each child, containing examples of work in different aspects of the curriculum.

The feeder infant school used the infant level of the same textbook series and the junior school staff saw this as a distinct advantage. As well as receiving records of achievement the lower junior teachers visited the infant school. Consequently they felt they had a good idea of their pupils' infant school experiences.

The head of the mathematics department in the comprehensive school to which virtually all the pupils transferred, visited the older junior pupils on several occasions and did some mathematics with them. This gave him a fair idea of the range of attainment of his first-year pupils.

Record keeping may be crucial to some mathematics schemes, especially those in which there is an emphasis on individualized learning:

In a secondary school which used an individualized learning scheme, the teachers of mathematics attached great importance to keeping comprehensive records for each pupil:

> as an aid to continuity;
> to monitor pupils' progress; and
> to set suitable work.

The teacher set a group of tasks, usually ten, for each pupil. After the pupils had completed all their tasks, they did a test in which each question was related to one of the tasks. The pupils were allowed to use their notebooks as well as any other available material to complete the test. The teacher then marked the tests, pointed to mistakes and asked the pupils to look at them again. After that they were re-marked.

For each group of tasks which the pupil completed, the records showed:

> the work set;
> the average level of difficulty of the set work;
> the time taken to complete the work;
> a mark for the test which was based on the given tasks;
> a second mark for the test after the pupil had attempted to correct mistakes pointed out by the teacher; and
> comments by the teacher on the work or the test.

The results of the test and the teacher's observations were used to set new work.

For some older pupils the records were used to assess their performance over a period of time and this information formed part of the assessment for a Mode III CSE.

Because low attainers often work more slowly and are behind the other pupils, it is more important for them that time at school is not wasted, so it is vital that there is continuity in their learning when they move to a new teacher or a new school. As well as recording attainment and experience, records for low attainers should include details of particular learning difficulties and physical, social or environmental factors which affect their ability to learn. For all pupils, but especially for low attainers, it should always be borne in mind that the major purpose of assessment is to help teachers provide learning experiences which are appropriate, particularly in regard to:

a pupil's present knowledge;
the difficulty of the concepts introduced;
sources of motivation;
present and future needs; and
style of presentation.

Points for discussion

1 What reasons for assessing low attainers in mathematics do you think are the most important?

2 What methods do you think are suitable?

3 Is selection necessary? If so, how can this be done? Is there provision for reviewing the allocation of pupils to teaching groups? In general do you think that your selection procedure identifies the right pupils?

4 When you assess pupils do you always use the information gained in order to modify their courses?

5 How and what do you record? Do the details of the records match the use which is made of them? Do you send/receive records to/from other teachers or schools? If so, what liaison is there?

6 If the pupils are given a grade or mark as a result of a public or an internal assessment, does it reflect positively what they have achieved?

References

1. W. E. C. Gillham and K. A. Hesse, *Leicester Number Test.* Hodder and Stoughton Educational, 1970.

2. S M I L E (Secondary Mathematics Individualized Learning Experiment). Inner London Education Authority, 1972– . Materials from I L E A Learning Materials Service, Highbury Station Road, London N1 1SB.

3. National Foundation for Educational Research, *Basic Mathematics Tests.* N F E R Publishing and N F E R–Nelson, 1969–80.

4. Gillham and Hesse, *Leicester Number Test.*

5. Science Research Associates, *Diagnosis: an Instructional Aid – Mathematics Level A* and *Mathematics Level B.* S R A, 1980–81.

6. L. M. Bental, *Essential Mathematics.* N F E R Publishing, 1976.

7. Inner London Education Authority, *Checkpoints Assessment Cards*, Primary Pack and Lower Primary Pack. I L E A, 1979. Materials available from I L E A Learning Materials Service, Highbury Station Road, London N1 1SB.

8. For example, M. L. Brown, *C S M S Number Operations.* N F E R Publishing, 1978. Also, *CSMS Diagnostic Maths Tests 1–10.* N F E R–Nelson, forthcoming.

9. ILEA Schools' Psychological Service, *Classroom Observation Procedure*. ILEA, 1979. Details available from Hoxton House, 34 Hoxton Street, London N1.

10. *The Kent Mathematics Project*. Ward Lock Educational for the Schools Council, 1978– .

11. H. Fletcher, A. Howell and R. Walker, *Mathematics for Schools*. Addison-Wesley, 1970–71.

12. Bental, *Essential Mathematics*.

13. See reference 8.

14. Inner London Education Authority, *Checkpoints Assessment Cards*.

15. See reference 8.

16. Nuffield Mathematics Project, *Checking Up 1, 2 and 3*. W. & R. Chambers and John Murray, 1970–73.

17. Inner London Education Authority, *Checkpoints Assessment Cards*.

18. Department of Education and Science, Assessment of Performance Unit, *Mathematical Development: Primary Survey Report Nos. 1 and 2*. HMSO, 1980, 1981. Also, Department of Education and Science, Assessment of Performance Unit, *Mathematical Development: Secondary Survey Report Nos. 1 and 2*. HMSO, 1980, 1981.

19. A. N. Hieronymus, E. F. Lindquist and N. France, *Richmond Tests of Basic Skills*. Thomas Nelson, 1975.

VII. Making improvements

Is there a need to change?

Two recent national reports[1] (referred to on pages 11 and 35–6) have highlighted the fact that the teaching of mathematics to low attainers is a matter of major concern and that considerable changes are required if the present situation is to be improved.

Teachers at most of the schools visited – all of which were recommended by local authorities – felt that they had not found a satisfactory solution to the problem of teaching mathematics to low attainers. Many of these schools were making and planning changes; some seemed to be monitoring and appraising their own work in the light of these changes.

Although there are no easy answers to the problems of teaching mathematics to low attainers this does not imply that improvement is impossible. One of the difficulties frequently encountered by teachers in attempting improvement, even in the schools visited, was that some colleagues were reluctant to change. Sometimes teachers agree about desirable changes, but these prove administratively difficult.

Some will resist change, particularly those teachers who lack confidence in themselves and underestimate their own abilities and expertise, and it may be difficult to convince others that particular changes will lead to a measurable improvement. In addition change usually involves extra work for teachers who are already under many short-term pressures and consequently lack time, especially time to give to an area of work which has had such low priority.

Real improvements in the provision for low attainers in mathematics can only happen when the individual recognizes and responds to the need for change. People can be compelled to change but, although this may result in new outward behaviour, it will not usually generate the attitudes that will ensure success.

At what levels can support for change be provided?

Even if extra resources of time and money, desirable as they are, are not available in a particular school, some improvements may be made by the reallocation and more appropriate use of existing resources. But it is unrealistic, impractical and inefficient to expect teachers to solve the complex problems of teaching mathematics to low attainers and institute the necessary change without considerable support at several different levels.

At a national or regional level, funds, which would be negligible in comparison with the cost of running a single secondary school, need to be made available for major investigations into the areas outlined in Chapter VIII ('Future needs'), principally:

1　the development of materials for use in in-service training
2　an examination of teaching approaches, teaching materials and their effectiveness in helping pupils to learn.

Some of this work could be undertaken at *local education authority* level; indeed, this could be the most appropriate level because many issues need to be considered in relation to local circumstances.

Essentially each authority should develop, as a high priority, a policy for low attainers in general and for low attainers in mathematics in particular, which should include:

encouraging headteachers and others to consider the education for low attainers in mathematics as a matter of importance;
providing support and in-service training for the whole staff of each school, individual departments in secondary schools and individual teachers; and
obtaining, displaying and giving advice about available materials – textbooks, workcards, materials, apparatus, etc.

To achieve this, advisers with specific responsibility for the work will need to be identified or appointed. They will require sufficient resources of time and money to fulfil the task adequately. In some cases advisers need the opportunity to acquire the necessary expertise.

At school level the priority is to develop a policy for low attainers in mathematics which is related to the school's policy for mathematics and the policy for low attainers. Generally, if the school has no overall policy for mathematics or for low attainers these will need to be developed first.

A policy relating to low attainers should include:

aims
teaching approach
staffing
curriculum
organization
resources
assessment, diagnosis and related action
records
continuity
in-service training

and, in secondary schools:

examinations and school-leaving certificate
work/school interface.

Most of these points should be considered by the school as a whole, and not only by the mathematics or remedial department in the secondary school or individual teachers in the primary school, since decisions may well affect and be affected by other aspects of the school. *Within the school someone should have responsibility for the teaching of mathematics to low attainers throughout the school.* At primary level this may be the head, deputy or a mathematics or remedial specialist who has the full support of the head. At secondary level it could be the head or another member of the mathematics department or a member of the remedial department.

Each teacher of low attainers in mathematics should be encouraged to examine his or her teaching methods, selection of materials and selection of mathematical content in relation to school policy and the particular pupils as well as to develop his or her own professional and managerial skills by participating in suitable in-service activities. When there is no overall policy related to the teaching of mathematics to low attainers, changes may still be made by individual teachers within the classroom, but the effects of these may be limited because of their isolation and lack of support from the rest of the school.

At national, regional and local authority level as well as within the school, provision is needed for in-service training relating to the teaching of low attainers. At all these levels, the education of these pupils needs to be kept under review.

How can change be brought about?

The following examples show how changes have been made at different levels. Although not all of them are primarily concerned with low attainers in mathematics, these pupils were directly affected by the change in every case.

AT NATIONAL OR REGIONAL LEVEL

A lead from the Inspectorate

Some years ago HMI, recognizing that there were major problems in teaching mathematics to slow learners in secondary schools and that there was little provision for in-service training, arranged a number of residential courses for teachers and advisers which were held annually. These have now led to Department of Education and Science regional courses which are run by individuals within the regions and attended by local mathematics advisers with groups of selected teachers for their areas, thus allowing discussion to be related to local needs. Attempts by schools or individuals to implement new ideas can be followed up by advisers in the area and form part of a continuing in-service education programme.

AT AUTHORITY LEVEL

Identifying a common need and encouraging cooperative work

In response to requests from a number of mathematics departments who were teaching or wanting to teach pupils in mixed-ability groups, a large LEA set up an organization to bring groups of teachers together on a regular basis to plan and produce individualized pupil material. Later a teacher was seconded full-time to run this group. As the scheme grew it was felt that there was insufficient material for low-attaining pupils, and a working group was formed to fill this gap.

Using advisory teachers to support work in schools

In one authority two primary school teachers with particular interest and expertise in teaching mathematics were seconded to a mathematics support team, initially for one year. They worked in schools which expressed concern or requested advice about different aspects of teaching mathematics. Much of their time was spent in helping teachers to identify difficulties in teaching mathematics to low attainers, plan work which would be appropriate for individual pupils and prepare suitable teaching materials.

Setting up a working group

In another authority, a mathematics adviser who was concerned about low attainers in mathematics drew together an interested group of teachers who spent time discussing the difficulties. They published a report which was subsequently widely circulated. This included a list of available materials and their usefulness, comments about organization of the teaching and a résumé of the causes of the learning difficulties experienced by low attainers in mathematics.

Introducing assessment materials developed by the local authority
An authority supported the work in primary mathematics by publishing a series of guidelines for teachers and materials for the assessment of pupils. In one infant school where these materials, *Checkpoints Cards*,[2] had been bought by the head teacher, two members of staff attended a course relating to the use of the cards. It was authority policy to run such courses and to encourage staff of schools which purchased the material to attend them. The teacher with responsibility for mathematics felt that use of the cards had provided a valuable starting point for a discussion of mathematics teaching throughout the school. She said that they were invaluable in developing teachers' awareness of pupils' difficulties. In her classroom, use of the *Checkpoints Cards* had led to a more individual approach to the teaching of mathematics.

AT SCHOOL LEVEL
Local authority advisers can influence the school by contact with teachers at all levels. It is not usually very effective to use one class teacher from a school as a way of influencing the whole school. First, the teacher may not be able to effect changes in his or her own teaching without additional support. Second, if the teacher does succeed in this, others may not feel able to follow the example, so this may lead to differing approaches and attitudes and potential professional isolation of the teacher. Usually it is more effective to influence the school through contact with the head, deputy head, or head of department. This may not lead to such dramatic changes in the short term, but it is likely to be more valuable because the whole staff or all the members of the department will be involved with all the changes that are agreed. In the primary school the teacher with responsibility for remedial work or mathematics can influence the school as a whole only if he or she has the active support of the head.

While some schools are seen to make dramatic changes in the teaching of mathematics to low attainers, these have, inevitably, been preceded by a considerable amount of work including detailed discussion, examination and selection of new materials. However, the school, department or teacher who is becoming aware of the need to change and wishes to respond will usually want to start with small but realistic changes. The problem is where to begin. The following examples illustrate some of the ways in which schools, departments and individual teachers have effected changes, often small, but nevertheless significant.

Introduction of new teaching approaches

In one first school which was open-plan, the headmistress, a member of the Association of Teachers of Mathematics, took a great interest in mathematics and wished to promote a particular way of working in which practical work and discussion were given a high priority. She felt that many of the difficulties pupils have with the subject stem from too much formal arithmetic, written down with little understanding, and a lack of practical work and discussion, particularly in the early stages. In order to implement her ideas she spent time working with pupils alongside their teachers. This led the staff to develop new teaching techniques: in particular there was much more discussion and questioning of pupils and informal recording of work by pupils. Changes were supported by the school's close contact with a lecturer in a local college of education. Student teachers visited and worked in the school; together with teachers they organized projects, including a study of 'the seashore'.

Keeping current practice under review

In an infant school, the staff met regularly over lunch once each week to discuss educational matters, such as policy about parts of the curriculum, concern about a child's or group of children's learning or how and when to teach a particular point. Visitors were invited to staff meetings from time to time; once an LEA mathematics adviser came and after a brief introductory talk from him they discussed problems related to the teaching of mathematics in their school. This encouraged further discussion, stimulated interest and led to changes in teaching.

Helping individual teachers to develop their skills

In another infant school in a very deprived urban area, there was a teacher with a post of responsibility for mathematics and one with responsibility for reading and language development, both of whom had attended LEA courses on 'slow learners' and had a particular interest in children's learning difficulties. The headmistress took occasional classes so that the teachers with special responsibility could go into the probationary teacher's classroom and work with her for a while, taking in some of the materials which they had used and found successful. The head also took the probationer's class so that she could visit other rooms and see how teachers organized their classes, used materials and taught particular aspects of the curriculum. The headmistress hoped to extend this practice to involve the whole staff.

Stimulating interest in mathematics

In one comprehensive school in a deprived area the head of the mathematics department organized a mathematics week. As well as organizing film shows, a poster exhibition, games, clubs and puzzle trails, he invited local mathematics advisers, college lecturers and people who were well known in mathematical education to visit the school and teach one or two classes. Every mathematics class in the school, including the low attainers, had a 'special lesson' with one of the visitors: the regular class teacher observed or joined in the lesson. Pupils, teachers and visitors all gained from the experience and a great deal of interest and enthusiasm for the subject was generated within the school.

Introduction of practical work
In a secondary school the teacher responsible for mathematics in the first, second and third years, and a colleague, wished to introduce practical work, particularly to pupils in the lower sets. Since little money was available and teachers had minimal experience of working with apparatus, a mathematics club was started, initially for first- and second-year pupils. Games, puzzles and other mathematical apparatus, many of them home-made, were tried out. In time more money became available and the amount and range of apparatus was increased. Staff and pupils alike benefited from being able to experiment and talk about materials in an informal, relaxed atmosphere. The experiences gained were invaluable in introducing the materials into lessons.

Encouraging and supporting staff in adopting a more individualized approach to teaching mathematics
In an urban comprehensive the remedial department taught all the pupils in the bottom band. The head of the remedial department wanted staff in his department to adopt a more individual approach to the teaching of mathematics, so he introduced an individualized learning scheme consisting of a large material bank from which the teacher selected appropriate material for each pupil. First he equipped his own room and, in using the scheme with his own classes, he solved some of the problems of organization and administration. Interested remedial teachers were then invited to team-teach with him. Only when they were familiar with the materials and the organization did he ask them to use the scheme with their own pupils, initially just with the first-years. He worked alongside the teachers until they felt confident. Later the scheme was extended to include second- and third-year pupils as well as other teachers.

Selecting an examination course to motivate older secondary pupils
The head of the remedial department in a boys' comprehensive school had taken a fourth-year bottom band group for a year, following a traditional Mode I CSE course. He was determined not to repeat the experience. The boys, he felt, had learned little, chiefly because they had no interest in the subject, but also because they had an inadequate understanding of the fundamental concepts which were necessary to cope with many of the topics on the syllabus. Before starting the next year's course, he looked at various materials and courses and decided that one particular Mode III CSE course used by another school which involved project work would be more suitable. He also examined available materials and decided to use some of the packs from the Schools Council's Mathematics for the Majority Continuation Project[3] despite reservations about whether his pupils would understand the written instructions.

Administration and matching work to pupils' interests and attainment was made simpler by restricting a pupil's choice of projects. They were set four assignments and allowed to choose the fifth. Initially pupils frequently requested assistance and sought reassurance while tackling their work, but as they gained confidence this decreased. The availability of electronic calculators allowed pupils to focus on solving the problems. Although some pupils preferred to work on their own, most worked in small groups tackling a common task. Those that did gained considerably from the discussion associated with cooperative working.

After a term's work various changes were apparent – a happy, relaxed atmosphere, fewer discipline problems and a marked fall in truancy. Interest in the work being done was expressed by other pupils as well as teachers, so it was decided to combine two groups with their two teachers to teach them as a team. Further assistance for pupils who needed help was given by two parents who came into classes on a regular basis.

Initiating practical work with older pupils

In a comprehensive school the second in the mathematics department was concerned because she felt that many pupils, in particular the low attainers, were being asked to work at a level of abstraction which was generally beyond them. To counter this she thought that a greater emphasis should be placed on practical work and materials-based learning, and she decided to set up a mathematics workshop in part of her large classroom. In addition to collecting a range of apparatus, mathematical games and puzzles, she also borrowed two programmable calculators from a local firm and was starting a collection of 'interest' books for a mathematics library. She hoped that all pupils in the fourth and fifth years would use the workshop. A mathematics club which would use all the facilities of the workshop and be open to all upper-school pupils was planned to start in the following term.

Trying out new ideas in an informal way

A fourth-year junior school teacher had attended a course where she had been introduced to a variety of new ideas, materials and methods, including Dienes Base 10 material. Whilst recognizing the potential value to pupils, she felt unable to introduce it into her classroom straightaway. She therefore organized a lunchtime activity for those who were having difficulty in mathematics in which she could try out some work with apparatus borrowed from the teachers' centre. After a while she felt sufficiently confident in using the material and in its educational value to use it in lesson time and the head allowed her to buy some of the apparatus.

Getting help by visiting teachers in other schools

In one junior school where most of the classes were run fairly formally and pupils were nearly always class-taught, one teacher, who felt that this approach was often not appropriate for her mixed-ability class and in particular the low-attaining pupils, tried to organize group work but found this very difficult. She talked to her headmistress who offered some advice and also made enquiries to see if any other teachers in the authority in a similar situation had been able to make an effective change. As a result the headmistress taught the teacher's class while she visited another school to observe a class whose teacher ran an integrated day. Discussion between the two teachers identified initial changes which could be made to introduce group work. They thought, for example, that for half an hour each day the teacher could work with one group of pupils whilst the rest of the class worked on a set task and read a book if they finished or 'got stuck'. It was thought to be important to make sure that every child in the class participated regularly in the small group work.

Changing teaching materials

A teacher who joined the remedial department of a comprehensive school took the first- and second-year bottom sets for mathematics. The head of the mathematics department gave her a set of textbooks for use with these pupils; the same text was used by children in other sets. After attempting to use this book for the first half-term the teacher felt that it was unsuitable and that the pupils' worsening behaviour and increasing dislike of mathematics was largely a result of the unrealistic demands which were made on them. After repeatedly expressing these opinions, fairly forcefully, to the head of the mathematics department she was eventually allowed to abandon the textbook and use her own worksheets and homemade apparatus. Later the head of department recognized that her approach was 'probably more suitable for bottom-set pupils' and allowed her some money for materials.

Changing from class to group work

A teacher with a third-year junior class was particularly interested in natural history. On a course run by the local authority he picked up many ideas to try out with his pupils, involving visits and practical work in the classroom. Although previously he had only class-taught, he found that practical work such as building a wormery or examining things under a microscope could only be done by a small group. This was partly because only a small group could work cooperatively, discuss plans and involve each individual; partly because the work initially needed careful supervision; but also because they had only a few microscopes. He divided the class into groups for science; each group did practical work in turn while the rest of the class did English. After the project had finished he reverted to class teaching for a while, but on reflection he thought that the group work had been successful, particularly for the children who found their school work difficult and were in consequence poorly motivated. Gradually he introduced more group work and more practical work into the classroom in science, mathematics and craft. He thought that most pupils had benefited from the change but that the most marked improvement was in the interest and enthusiasm shown by the lowest attainers.

Stimulating interest by having a guest speaker

The teacher of a third-year class in a middle school invited a friend, a civil engineer, to talk to her class about his work building and designing roads, bridges and buildings. After a brief talk the pupils asked a great number of questions: 'How much did it cost to build our school?' 'How long did it take?' 'How much do engineers get paid?' 'How much do builders get paid?' 'How do you put in the electricity?' 'Who pays for all the buildings?' 'Where do they get the money from?' The interest of the pupils including the low attainers led to further work about both building and money – wages, salaries, banking and public funds.

Getting started

JOINING OR FORMING A GROUP

Many teachers belong to national associations which take some interest in the teaching of mathematics to low attainers (see Appendix A for details).

LOOKING AT THE EFFECTIVENESS OF PRESENT SCHEMES AND WAYS OF WORKING

Several authorities have designed self-assessment schemes for schools, departments and teachers while others have asked all the schools in their authority to produce detailed policy documents. Both actions have led schools to take a closer look at their work. Sometimes schools themselves have initiated similar exercises.

It is imperative that changes be realistic in terms of the demands made upon individuals, and that the details be very carefully planned. Problems and difficulties can be minimized by communicating with others who are contemplating or have attempted similar changes; advisers can play a key role by putting schools in touch with each other to facilitate this interchange.

Points for discussion

1 a Which features of the teaching of mathematics to low attainers in your school are most in need of improvement?
   b Are there small and realistic ways in which you could change the present practice to improve matters?
2 Which aspects of the problem need to be tackled:
   a by the mathematics or remedial departments?
   b at school level?
   How can this best be done?
3 What action if any, do you think should be taken by your LEA to assist schools in teaching low attainers in mathematics adequately? Have these needs been expressed to those who are responsible?

References

1. W. K. Brennan, *Curricular Needs of Slow Learners* (Schools Council Working Paper 63). Evans/Methuen Educational, 1979. Also, Department of Education and Science, *Aspects of Secondary Education in England*, a survey by HM Inspectors of Schools. HMSO, 1979.
2. Inner London Education Authority, *Checkpoints Assessment Cards*, Primary Pack and Lower Primary Pack. ILEA, 1979. Materials available from ILEA Learning Materials Service, Highbury Station Road, London N1 1SB.
3. Schools Council, *Mathematics for the Majority Continuation Project* (packs 1, 2, 3 and 4). Schofield and Sims, 1974–75.

VIII. Future needs

During the school visits, the project team asked many teachers to suggest further work which would help them in their teaching of low attainers in mathematics.

A number of teachers stressed that any work should take into account the realities of teaching, in particular teachers' time and skills and the limited funds available. Many were also concerned that any work should not ignore the requirements of other aspects of education for the limited resources available. The priorities for further work, as teachers saw them, included:

1 smaller classes which would make it easier to provide practical work and give more individual attention;
2 an increase in non-teaching time to allow more opportunities for teachers to discuss problems and review current practice;
3 improved initial and in-service training in:

> the early stages of mathematical development,
> effective teaching techniques,
> the selection and appropriate use of materials,
> techniques of organization and control; and

4 more effective ways of identifying the point at which the pupil fails, and suggestions for appropriate remedial action.

Recommendations for further work

In the light of these comments and the experiences of the steering committee and project team, it seems that the needs for further work fall into two categories:

1 *Teaching training*: collecting information which is relevant to the teaching of low attainers and providing teachers with appropriate access to the material so that it will be of practical value.

2 *Research into teaching and learning mathematics* and using the results to assist in the development of teaching and learning materials.

Alongside these is a continuing need for development and refinement of teaching material arising from activities which have been found to work well in the classroom. This need is at all levels: from particular activities devised by the individual teacher to complete published schemes.

TEACHER TRAINING

In teaching low attainers in mathematics, knowledge and skills are required in two different fields:

the teaching of children with particular problems; and
the early stages of conceptual development in mathematics, the applications in real life of very basic mathematical concepts, and understanding of the way that the early stages relate to later stages of mathematics.

In both these fields an awareness is needed of the theories which have been put forward by educational psychologists and their implications in practical terms, as well as advice about organization and control in the classroom; a group of low attainers will almost certainly contain some disruptive pupils.

The preparation of materials such as videotapes of lessons, recordings of discussion between pupils, or between teacher and pupil, and case histories of low attainers in mathematics, would be of great value to a teacher-training programme. Similarly those involved in teacher training and teachers themselves would be helped if reference material were compiled containing, for example:

information relating to materials;
ideas about how topics and projects could be presented; and
suggestions about the different stages of development within each topic.

There is now a wide range of pupils' materials available. What is required is that teachers be given the opportunity to examine the variety of resources and be helped to assess them so that they can select and use what is appropriate for their pupils. Even so, there are specific areas in which further development is needed, for example materials which build on real-life applications at all levels, from infant school upwards.

Although a start can be made in many of these areas in initial training, it must be recognized that most of the work can only realistically be tackled once teachers are in post, and have a background of relevant experience on which to build.

RESEARCH INTO TEACHING AND LEARNING MATHEMATICS

As well as making present knowledge available and useful to teachers, there are several areas in which little is known and additional research should be undertaken. The priorities for work are as follows:

1　*Teaching and learning mathematics:*
examination of different approaches to the teaching of mathematics to determine their effectiveness;
careful documentation of work done with a group of low attainers, including the different tactics used and the pupils' responses over a period of time in order to gain insight in relating specific techniques to particular pupils;
examination of different learning strategies adopted by pupils: how are successful strategies developed and what is it that ensures their success?

2　*Assessment:*
development of screening techniques to identify pupils at risk;
investigation and development of ways to determine a pupil's present stage of development in different mathematical topics, and to link the results of this assessment to 'learning pathways' by recommending appropriate work and teaching approaches.

3　*Mathematical needs:*
investigation to determine the types of mathematical learning which pupils are likely to need at school and in adult life, both at home and at work.

4　*Curriculum development* based on the research outlined above.

It is hoped that it will be possible to make a start on this work in the near future.

Appendices

Appendix A Sources and resources

Sources of information

The main sources of information are people, organizations and publications.

PEOPLE

Teachers
One of the most valuable sources of help are other teachers who have had relevant experience. Information or advice may be available from other staff in the school or from teachers in other schools.

Advisers
Local authority advisers can assist an interchange of views by putting teachers in different schools in touch with one another. Lecturers in local colleges of education as well as the authority's mathematics, remedial and primary advisers may be able to inform, both through in-service courses and in response to a direct approach.

ORGANIZATIONS
There are national organizations which publish books, reports and periodicals containing material of interest to the teachers of low attainers in mathematics. The periodicals also include reviews of materials. Each organization has an annual national conference and affiliated branches that meet locally. Those most likely to help the teachers of low attainers in mathematics are:

The Mathematical Association
259 London Road
Leicester LE2 3BE
Periodical (published five times yearly): *Mathematics in School.*

The Association of Teachers of Mathematics
Market Street Chambers
Nelson
Lancashire BB9 7LN
Periodical (published quarterly): *Mathematics Teaching*.

The National Association for Remedial Education
2 Lichfield Road
Stafford, ST17 4JX
Periodical (published quarterly): *Remedial Education*.

OTHER PUBLICATIONS

Two articles in *Mathematics in School* summarize, for the age groups
14–16 and 9–13, teachers' opinions of materials which are commercially
available, including books, apparatus and games. They are:

for 14- to 16-year-olds: D. Lumb, 'Mathematics for the less gifted',
Mathematics in School, 7(2), March 1978; and
for 9- to 13-year-olds: D. Lumb, 'Mathematics and the less gifted child
in the middle years', *Mathematics in School*, 9(3), May 1980.

Newsletters which review books, games, apparatus, tests, etc. and
comment extensively on materials and the teaching of low attainers in
mathematics in general are published by:

Scottish Curriculum Development Service
Dundee Centre
College of Education
Gardyne Road
Broughty Ferry
Dundee DD5 1NY, Scotland

They are called *Mathematics for the Least Able 11–14*,*but much of the
information is also valuable for teachers of pupils outside this age range.
Published once per term.

Struggle is a magazine produced (twice yearly) by the Inner London
Education Authority, first issued in 1979. It is concerned with the teaching
of mathematics to low attainers of 11 years and over, including those in
special schools, and contains various articles, reviews of materials and
suggestions for work. *Struggle: Mathematics for Low Attainers* is available
from:

---

* Published as a service to Scottish schools; also sold to a small number outside of Scotland.

Hackney Teachers' Centre
Digby Road
Homerton
London E9 6HX

Resources

During the school visits, many resources, both commercial and teacher-produced, were brought to the attention of the project team. These could be categorized under the following headings:

GENERAL INFORMATION
Local authority guidelines/syllabuses.
Relevant local authority documents, surveys, etc.
Teachers' centre handouts.
CSE courses/authority non-examination courses/leaving certificates.
National reports: HMI's, NFER's, Schools Council's.

WRITTEN MATERIAL
Books and articles written for teachers.
Pupil materials: schemes, workbooks, workcards, worksheets, general interest/topic books, etc.

ASSESSMENT MATERIALS
Tests.
Assessment tasks.
Observation procedures.

APPARATUS
For: number work – structured and non-structured,
        measurement,
        probability and statistics,
        general information: maps, timetables, brochures, etc.,
        shape,
        sets and logic,
        mathematical paper: graph, isometric, etc.,
        games and puzzles,
        domestic and natural materials.

AUDIO-VISUAL MATERIALS
Radio/TV programmes, both local and national
Tapes
Films

FURTHER INFORMATION

A detailed list of materials can be extremely valuable if it contains enough specific information to allow the user to make appropriate choices or if items are selected according to stated criteria which match those of the user. Difficulties became apparent when the project team considered writing such a list for this report. First, any list compiled at the time of writing would be out-of-date by the time this book is published, because new resources would be available and others would have been withdrawn. It was thought better to suggest the above sources, from which up-to-date information can be obtained.

Second, any list would be very long, because it would need to include general educational material as well as that which is produced specifically for low attainers, if it is to be applicable to the whole age range. Some of the above materials might be used selectively to satisfy a very specific need. Any shortened list would imply some evaluation by the project team and insufficient time was available to make a detailed survey and review of materials.

Third, whilst recognizing that a range of suitable resources will help the teacher, it is felt that no text can be entirely suitable for low attainers, since questioning, discussion and mental work are seen to be of great importance.

Fourth, most teachers are likely to have access to collections of suitable material held at teachers' centres, where advice may be available.

After considering these points it was felt that in the time available to the team it would be difficult to improve on references already made and the many lists compiled by local groups of teachers recommending materials related to local circumstances. However, the report by members of this group engaged in the literature review (see Introduction, page 12) contains an extensive bibliography as well as reviews of relevant research literature, including the use of particular materials.

Appendix B   Features contributing to success

As a result of the school visits, the project team and steering committee identified a number of features which they felt contributed to the success of the work with low attainers in mathematics in schools.

It is stressed that this list is a subjective one. Some of the points overlap, and there is no claim that the list is comprehensive. The features identified were:

1   There should be a planned curriculum related to the pupil's immediate and future needs, including aesthetic appreciation.
2   Pupils should start from their own level of achievement rather than from a common point of failure, which implies some form of individual assessment.
3   Work should be well matched to the pupil and presented in a way that will facilitate successful learning.
4   The presentation of work should emphasize: oral work and discussion; variety in the presentation and communication of mathematical ideas; and approximation and estimation.
5   Teaching should be essentially material based and where appropriate realistic applications should be used.
6   Teaching should include the sensible use of electronic calculators and these should be easily accessible to pupils.
7   Where there is withdrawal or a remedial group, the organization should encourage transfer back to the mainstream when appropriate.
8   The programme of work should be flexible enough to take into account pupils' interests.
9   Pupils should understand what they are doing and why, and should enjoy working.
10   There should be a wide variety of approaches and stimuli.
11   The curriculum should be wide and not restricted to computation.
12   There should be a conscious objective of helping pupils to organize their own work and become self-reliant.
13   Wherever possible mathematics should be developed across the curriculum.

Steering committee and project team

Steering committee

J. W. Hersee (Chairman) — Executive Director, School Mathematics Project and Chairman of Schools Council Mathematics Committee

J. Bancroft — Head of Mathematics Department, Westfield School, Bedford (now retired)

R. Berrill — Lecturer in Education (Mathematics), School of Education, University of Newcastle-upon-Tyne

J. Brand — Warden, Hipper Teachers' Centre, Chesterfield, Derbyshire

M. Cahill — Deputy Headmaster, Frank F. Harrison School, Walsall

N. Cawley — Lecturer in Special Education, Avery Hill College, London

D. L. Davies — Headmaster, Christopher Whitehead Boys' Secondary School, Worcester

R. Edwards — Adviser for Remedial and Special Education, Knowsley

E. M. Hardwick — Adviser for Nursery and First Schools, Buckinghamshire

R. W. Hart — Senior Inspector for Special Education, Walsall

B. L. Hensman — Teacher responsible for Remedial Mathematics, Highfield Middle School, Keighley (now retired)

D. M. Jones — HM Inspectorate of Schools

G. Knights	Director of Mathematics and Tutorial Faculty, Redhill Comprehensive School, Arnold, Nottinghamshire (until 1980); Deputy Head, Broadland High School, Hoveton, Norfolk (from 1980)
E. Parkes	Headmaster, Gusford Primary School, Ipswich, Suffolk
M. Ward	National Foundation for Educational Research
K. Wimpenny	Deputy Head, The Denes High School, Lowestoft, Suffolk

Project team

M. Brown (honorary director)	Lecturer in Mathematics Education, Chelsea College, University of London
I. Provan	Secretary

General report

C. Stolz (deputy director)	Philippa Fawcett Mathematics Centre, Inner London Education Authority (seconded)
B. Denvir (primary case study officer)	London Borough of Haringey (seconded)

Literature review

L. Dickson	Chelsea College, University of London
O. Gibson	Chelsea College, University of London

Schools Council staff

J. A. Denyer	Curriculum Officer
D. Clemson	Research Officer

Other publications from
the Schools Council

Statistics in schools 11–16: a review
by Peter Holmes, Ramesh Kapadia and G. Neil Rubra, edited by Daphne Turner

This report of the Schools Council Statistical Education Project (11–16), set up at the University of Sheffield in 1975, is a summary of project papers produced by the team in assessing the current situation in statistical education in schools, and covers the use of statistics in many areas of the curriculum.

Chapter I describes a survey carried out by the project team in a 10 per cent sample of all secondary schools in England and Wales. It reveals what statistics is taught, by whom, to whom and in what subjects. Chapter II considers the statistical content of examinations at GCE O and AO levels and in CSE, investigating the incidence of topics and the types of examination questions asked. Chapter III looks more closely at the statistics pupils meet in their mathematics and science lessons, while Chapter IV similarly investigates the use of statistics in humanities and social science courses. In Chapter V the position of probability and statistics in the primary school is considered – providing a useful guide in assessing the statistical background which may be expected of pupils when they start secondary school.

The questionnaire used in the survey is reproduced in the appendices, which also include a review of some standard statistics texts and a comprehensive list of references for in-service courses, equipment, books and articles all relevant to the teaching of statistics.

The report provides useful information on the current position of statistics teaching in schools, and helpful and detailed reviews and lists of resources.

Methuen Educational, 1981. 144pp.

The practical curriculum

When the Schools Council decided its programme priorities in 1979, having consulted the local authorities, teachers and its other members, it considered that 'there appears to be support for developing a general framework for the curriculum. We propose to engage in discussions about curriculum balance and content. We believe that teachers, schools and local education authorities will welcome a lead from the Council.'

The practical curriculum presents the first fruits of those discussions. It argues that we need to develop a visible structure for the curriculum. A useful starting point is to consider each pupil's right of access to different areas of human knowledge and experience. For each child, what he or she takes away from school is the effective curriculum. This paper emphasizes this different view of the curriculum as the experience each child has at school and what each takes away.

The paper first discusses the principles which underpin a school's curriculum, and then considers the contribution to a child's education made by different kinds of experience and forms of knowledge, modes of teaching and learning, the development of values and attitudes, and the acquisition of skills, including non-verbal skills. Later chapters review the steps involved in planning and monitoring a curriculum and assessing a school's success in meeting its aims.

This contribution to the curriculum debate is addressed first of all to teachers. They are the people who have to resolve the tensions between the delivery of a mass public service open to all children and the individuality of each child. Whatever the outcome of the discussions on freedom and uniformity in the curriculum, many important decisions about what is taught and how it is taught will still be made by schools and by individual teachers. They need a good basis for these decisions, and that is what *The practical curriculum* provides.

Methuen Educational, 1981. 76pp.

Making the most of the short in-service course
by Jean Rudduck

Short in-service courses are the most common form of in-service training, yet there has until recently been little attempt to analyse their potential. Between 1979 and 1980 the Schools Council project, Making the Most of the Short In-Service Course, collaborated with five LEAs in studying local experience of various aspects and styles of in-service courses including short courses that meet once a week for several weeks, a programme of meetings organized by groups of first schools, teacher-tutor schemes, follow-up after a short course is over, and the evaluation of short courses. This report describes several such activities and approaches, draws some general conclusions about effective practice and raises issues for further debate.

Teachers, advisers and teachers' centre wardens alike will find this book invaluable in developing their own in-service courses.

Methuen Educational, 1981. 200pp.

Information skills in the secondary curriculum
edited by Michael Marland

This bulletin has been prepared by a working group sponsored by the British Library and the Schools Council. The recommendations will help middle and secondary schools include in their overall curriculum the development of pupils' information skills: the ability to formulate and focus a question, find possible sources, judge their appropriateness, extract the relevant information, reorganize it and prepare it for future use or presentation to others – in short, to encourage and assist students to organize their learning during school and beyond.

'It is a central responsibility of the school to help its pupils to cope with learning,' state the authors, and '... information skills can only be satisfactorily incorporated in the school's programme by a curriculum policy built around these skills'. The working group, by dissecting an apparently simple assignment, identifies a series of stages, nine 'question steps', which students need to work through. These stages are examined, and ways in which teachers might stimulate the development of the necessary study skills are discussed, together with their relationship to the overall curriculum.

Every teacher will be interested in this report because the general principles apply to learning tasks given to pupils and students of all ages.

Methuen Educational, 1981. 64pp.